Handbook of
Radiographic
Positioning
for Veterinary
Technicians

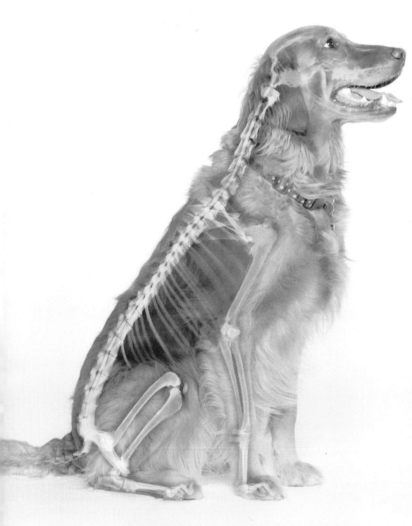

Margi Sirois, EdD, MS, RVT
Program Director of Veterinary
   Technology
Penn Foster College
Scottsdale, AZ

Elaine Anthony, MA, CVT
St. Petersburg College
St. Petersburg, FL

Danielle Mauragis, CVT
University of Florida
Gainesville, FL

DELMAR
CENGAGE Learning

Australia • Canada • Mexico • Singapore • Spain • United Kingdom • United States

DELMAR
CENGAGE Learning

Handbook of Radiographic Positioning for Veterinary Technicians
Margi Sirois, Danielle Mauragis, and Elaine Anthony

Vice President, Career and Professional Editorial: Dave Garza

Director of Learning Solutions: Matthew Kane

Acquisitions Editor: Benjamin Penner

Managing Editor: Marah Bellegarde

Senior Product Manager: Darcy M. Scelsi

Editorial Assistant: Scott Royael

Vice President, Career and Professional Marketing: Jennifer McAvey

Marketing Manager: Erin Brennan

Marketing Coordinator: John Sheehan

Production Director: Carolyn Miller

Production Manager: Andrew Crouth

Content Project Manager: Allyson Bozeth

Art Director: Dave Arsenault

Library of Congress Control Number: 2009905813

ISBN-13: 978-1-4354-2603-0
ISBN-10: 1-4354-2603-7

Delmar
5 Maxwell Drive
Clifton Park, NY 12065-2919
USA

Cengage Learning is a leading provider of customized learning solutions with office locations around the globe, including Singapore, the United Kingdom, Australia, Mexico, Brazil, and Japan. Locate your local office at: **international. cengage.com/region**

Cengage Learning products are represented in Canada by Nelson Education, Ltd.

To learn more about Delmar, visit **www.cengage.com/delmar**
Purchase any of our products at your local college store or at our preferred online store
www.ichapters.com

Notice to the Reader
Publisher does not warrant or guarantee any of the products described herein or perform any independent analysis in connection with any of the product information contained herein. Publisher does not assume, and expressly disclaims, any obligation to obtain and include information other than that provided to it by the manufacturer. The reader is expressly warned to consider and adopt all safety precautions that might be indicated by the activities described herein and to avoid all potential hazards. By following the instructions contained herein, the reader willingly assumes all risks in connection with such instructions. The publisher makes no representations or warranties of any kind, including but not limited to, the warranties of fitness for particular purpose or merchantability, nor are any such representations implied with respect to the material set forth herein, and the publisher takes no responsibility with respect to such material. The publisher shall not be liable for any special, consequential, or exemplary damages resulting, in whole or part, from the readers' use of, or reliance upon, this material.

Printed in the United States of America
2 3 4 5 6 7 8   16 15 14 13

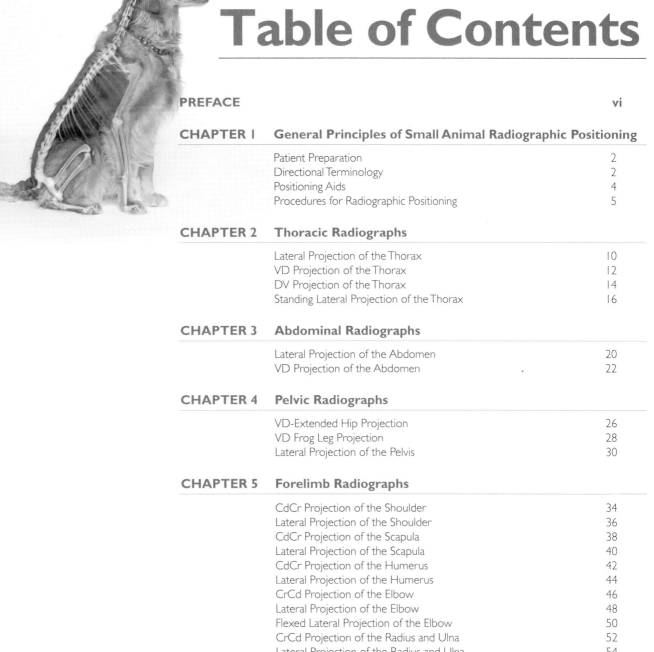

Table of Contents

Preface

Radiographic evaluation is a valuable diagnostic tool, and the veterinary technician plays a vital role in providing high-quality images for evaluation by the clinician. Proper patient positioning is crucial to achieving diagnostic quality images. This book provides detailed information on positioning of dogs, cats, birds, and pocket pets for radiographic examination. Photographs are used to illustrate correct patient positioning for each radiographic image. The resulting radiograph produced is also included, and diagrams are included for most images. Canine and feline dental radiographic techniques are also included. Although not meant to be a comprehensive radiology textbook, detailed information on patient positioning, positioning aids, and labeling of radiographic images is included. Positioning techniques described are most useful for clinical locations in which patients are sedated or anesthetized for radiographic procedures. However, all procedures can also be performed while manually restraining the patient if needed. Positioning techniques presented will provide diagnostic quality images when used with either traditional or digital radiology systems. Readers are encouraged to consult a comprehensive radiology textbook for additional information on production of x-rays, film processing, and safety issues related to the production of radiographs.

About the Authors

Dr. Sirois is the Program Director of Veterinary Technology at Penn Foster College. She received her AAS degree in veterinary technology from Camden County College, and also holds a BS and an MS in Biology and an EdD in instructional technology and distance education. She is certified as both a veterinary technician and laboratory animal technician, and has over 20 years' experience as a veterinary technician educator in both traditional and distance education programs. Dr. Sirois is a past president of the Association of Veterinary Technician Educators and a member of the editorial board for Veterinary Technician.

Elaine Anthony is an associate professor of veterinary technology in both the on-campus and online veterinary technology programs at St. Petersburg College, and is employed part-time at a veterinary internal medicine clinic. She holds an AS degree in veterinary technology from St. Petersburg College. She also holds an AA in music, a BS in elementary education, and MA in adult education. She is certified as a veterinary technician and has earned a certificate in Veterinary Hospital Management. She has over 20 years' experience as a veterinary technician educator. She is also a consultant for Nestle Purina and a member of the editorial board for Veterinary Technician.

Ms. Anthony and Dr. Sirois speak at several veterinary conferences each year, and have published numerous journal articles and textbook chapters on a variety of veterinary technology topics. Dr. Sirois has also authored and edited several veterinary technology textbooks, including *Principles and Practices of Veterinary Technology*, *Laboratory Animal Medicine: Principles and Procedures*, and *Laboratory Procedures for Veterinary Technicians*, all published by Elsevier.

Danielle Futch Mauragis received her A.S. degree in veterinary technology from St. Petersburg College. After working in private practice, she joined the Diagnostic Imaging service at the University of Florida Veterinary Medical Center. Her duties include x-ray, CT, Nuclear Medicine, assisting in ultrasound, and teaching physics, techniques, and quality control of radiology to veterinary students. Her hobbies are photography and sewing, but her passion is training dogs for

competition in agility. She shares her life with her husband Dennis, daughter Savannah, 2 dogs, Flik and Fizzy, and 1 ex-research cat name Pounce de Leon.

Contributors

Lori A. Barnes, CVT
Avian & Animal Hospital of Bardmoor, Largo, FL

Vickie Byard, CVT, VTS (Dentistry)
Rau Animal Hospital Glenside, PA

Deborah L. Walker, CVT
Avian & Animal Hospital of Bardmoor, Largo, FL

Acknowledgements

Special thanks to our families for putting up with us while we focused on this book. This book would not have been possible without your love and support. We are sending a special woof and extra dog treats to our favorite models, Dakota , Aspen, Woody, and Flik. We are grateful for the guidance and support of Senior Product Manager, Darcy Scelsi, and to our contributors for all their hard work.

Reviewers

Mary H. Ayers, BBA, RT(R)
Virginia Medical Regional College Of Veterinary Medicine
Blacksburg, VA

Karen Lee Hrapkiewicz, DVM
Wayne State University
Detroit, MI

Karl M. Peter, DVM
Foothill College
Los Altos Hills, CA

Lois Sargent, DVM
Miami Dade College
Miami, FL

P. Alleice Summers, DVM
Cedar Valley College
Lancaster, TX

Frances Turner, RVT
McLennan Community College
Waco, TX

Dedication

To our students—past, present, and future. You are our inspiration.

CHAPTER 1

GENERAL PRINCIPLES OF SMALL ANIMAL RADIOGRAPHIC POSITIONING

OVERVIEW

Positioning of small animal patients for radiography requires knowledge of normal anatomy of the species and descriptive directional terminology. When patients are not properly positioned, inaccurate interpretation of the radiograph and subsequent incorrect diagnosis of the patient can occur. Proper patient positioning usually requires that the patient be immobilized, either using **chemical restraint** (administration of sedatives and anesthetics) or **mechanical restraint** (use of radiology positioning aids). Manual restraint should be avoided due to the risk of increased exposure of the handler to ionizing radiation. When manual restraint is necessary, the handlers should take precautions to minimize exposure to ionizing radiation. This includes proper positioning, as well as moving as far as possible from the primary x-ray beam and wearing proper safety apparel, including lead-lined gloves, apron, thyroid shield, and glasses.

PATIENT PREPARATION

The veterinary technician should ensure that all animals being radiographed have a clean, dry haircoat. Wet hair and debris can cause confusing artifacts on radiographs. If possible, remove all collars and leashes, topical medications, bandages, and splints.

Animals undergoing radiographic study must be properly restrained. Ideally, chemical restraint with sedatives or anesthetics will be used so a handler does not have to remain in the room. This will also minimize the possibility of motion artifacts on the finished radiograph as well as minimize the anxiety of the animals. In some locations, laws prohibit manual restraint. When manual restraint is necessary, it should be accompanied by the use of positioning aids and the proper use of lead shielding to minimize exposure of the technician to radiation. The comfort of the patient must also be considered.

Careful planning and preparation will reduce the total time that patients must remain in position on the x-ray table. Identify all radiographic views needed, and prepare all supplies and equipment before placing the animal onto the x-ray table.

DIRECTIONAL TERMINOLOGY

A basic knowledge of directional terminology is required for proper patient positioning and for use when describing radiographic projections. The American College of Veterinary Radiology (ACVR) determines standard nomenclature for radiographic projections. The accepted nomenclature system requires that radiographic projections be named using only approved veterinary anatomical directional terms or their abbreviations. Radiographic projections are described using the directional term that describes the penetration by the central ray of the primary x-ray beam through the anatomical area of interest from the point of entrance to the point of exit. ACVR guidelines also include recommended nomenclature to be used when combinations of terms are needed and when oblique x-ray angles are used. In small animals, it is assumed that the primary x-ray beam is generated from a location above the animal unless otherwise indicated.

Commonly Used Directional Terminology:

Dorsoventral (DV): This term describes a radiograph produced when the primary x-ray beam enters the dorsal (topline or spinal) surface and exits the ventral (sternal or thorax and abdomen) surface of the patient (Figure 1-1a).

Ventrodorsal (VD): This term describes a radiograph produced when the primary x-ray beam enters the ventral

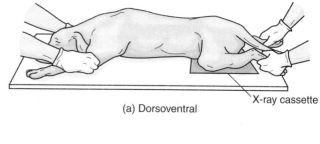

(a) Dorsoventral   X-ray cassette

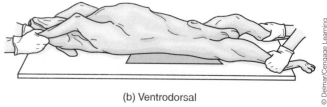

(b) Ventrodorsal

FIGURE 1-1

Dorsoventral versus ventrodorsal positioning.

surface and exits the dorsal surface of the patient (Figure 1-1b).

Medial (M): This term refers to the direction toward an animal's midline (Figure 1-2). The term is usually used in combination with other directional terms to describe oblique projections. For example, dorsomedial refers to the direction of the x-ray beam from the dorsal surface toward the midline. Radiographs of the limbs taken with the primary x-ray beam entering the medial surface of the limb and exiting the lateral may be referred to as mediolateral, although this is normally shortened to simply L.

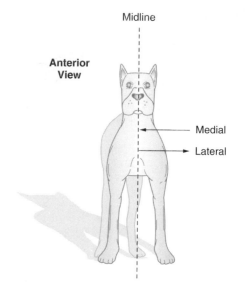

Midline

Anterior
View

Medial

Lateral

FIGURE 1-2

Medial versus lateral

© Delmar/Cengage Learning

Lateral: The term describes a radiograph produced when the primary x-ray beam enters from the side, away from the medial plane or midline of the patient's body. In the strictest use of ACVR nomenclature, a lateral projection taken with an animal lying on its right side would be referred to as left to right lateral. However, by convention, this is usually shortened to simply right lateral, meaning that the patient is positioned in right lateral recumbency, indicating that the patient's right side is closest to the x-ray cassette. Similarly, a limb radiograph obtained with the patient in right lateral recumbency with the affected limb placed against the x-ray table or cassette and the x-ray beam penetrating from the medial to lateral direction is referred to as a right lateral projection.

Proximal (Pr): This is a relative directional term that indicates a structure located closer to a point of attachment or origin from another structure or closer to the midline of the animal (Figure 1-3).

Distal (Di): This is a relative directional term that indicates a structure located farther from the point of attachment or origin of another structure or away from the midline of the animal (Figure 1-3).

Rostral: This relative directional term indicates a structure located closer to the nares from any point on the head (Figure 1-3).

Cranial (Cr): This relative directional term indicates a structure located closer to the animal's head from any part of the body (Figure 1-3).

Caudal (Cd): This relative directional term indicates a structure located closer to the animal's tail from any part of the body (Figure 1-3).

Plantar: This term is used to describe the caudal (posterior) surface of the hindlimb distal to the tarsus; the correct term for the surface proximal to the tarsus is caudal (Figure 1-3).

Palmar: This term is used to describe the caudal (posterior) surface of the forelimb distal to the carpus; the correct term for the surface proximal to the carpus is caudal (Figure 1-3).

Craniocaudal (CrCd): This term describes a radiographic projection obtained by passing the primary x-ray beam from the cranial surface to the caudal surface of a structure. It is most commonly used for radiographs involving the extremities proximal to the carpus or tarsus. Older veterinary literature may refer to this radiographic projection as anterior-posterior (AP).

Caudocranial (CdCr): This term describes a radiographic projection obtained by passing the primary x-ray beam from the caudal surface to the cranial surface of a structure. It is most commonly used for radiographs involving the extremities proximal to the carpus or tarsus. Older veterinary literature may refer to this radiographic projection as posterior-anterior (PA).

Dorsopalmar (Dpa): This term is used to describe radiographic views distal to the carpus obtained by passing the primary x-ray beam from the dorsal direction to the palmar surface of the forelimb. Older veterinary literature may refer to this radiographic projection as anterior-posterior (AP).

Palmar dorsal (PaD): This term is used to describe radiographic views distal to the carpus obtained by passing the primary x-ray beam from the palmar surface of the forelimb toward the dorsal surface of the body. Older veterinary literature may refer to this radiographic projection as posterior-anterior (PA).

Dorsoplantar (Dpl): This term is used to describe radiographic views distal to the tarsus obtained by passing the primary x-ray beam from the dorsal direction to the plantar surface of the hindlimb. Older veterinary literature may refer to this radiographic projection as anterior-posterior (AP).

Plantardorsal (PlD): This term is used to describe radiographic views distal to the tarsus obtained by passing the primary x-ray beam from the plantar surface of the forelimb toward the dorsal surface of the body. Older

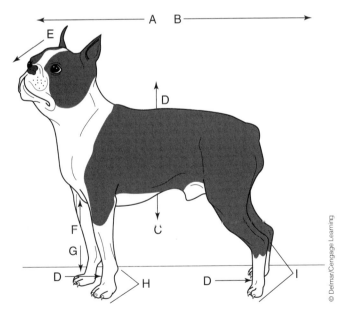

© Delmar/Cengage Learning

FIGURE 1-3

Common directional and positioning terminology. The arrows on this Boston terrier represent the following directional terms: A = cranial, B = caudal, C = ventral, D = dorsal, E = rostral, F = proximal, G = distal, H = palmar, I = plantar.

veterinary literature may refer to this radiographic projection as posterior-anterior (PA).

Oblique (O): This term refers to radiographic projections taken with the primary beam entering at an angle other than 90 degrees to the anatomical area of interest. Oblique projections are sometimes used to obtain images of structures that might be superimposed over other structures with standard 90-degree views. Nearly all dental radiographs are obtained using oblique angles. The angle used may vary depending on the site of interest. The specific angle should be included in the description of the radiograph along with the proper terminology to describe the direction of the primary beam. For example, a D60LMPaO indicates that the x-ray beam entered the dorsal surface at a 60-degree angle and exited at the medial area of the palmar aspect of the hindlimb. This nomenclature can become unwieldy and is often shortened for standardized oblique views used in a particular setting.

POSITIONING AIDS

Radiology positioning aids are used to increase the patient's comfort as well as ensure proper positioning for the radiographic evaluation. They may also allow for patient evaluation without a handler having to remain in the room. When manual restraint is needed, positioning aids will assist the handler in maintaining the animal in the correct position. Positioning aids should be small and lightweight to allow ease of use and storage. Most positioning aids leave some density shadows on the finished radiograph and should, therefore, not be placed over or under the area of interest. Positioning aids made of plastic are radiolucent, meaning x-rays can pass through the object. Reusable positioning aids must be waterproof, washable, and stain resistant. Reusable positioning devices include sandbags, foam pads and wedges, beanbags, troughs, and ropes. Disposable positioning aids include porous nonelastic tape, plastic or latex tubing, and roll gauze.

Sandbags

Sandbags can be purchased commercially (Figure 1-4), or they can be made from materials purchased at a craft store. Commercially prepared bags usually come prefilled with clean silica sand and are permanently sealed. They are commonly made of vinyl or nylon with plastic linings. Empty bags with sealable openings may also be purchased and filled with sand. Canvas bags cannot be easily disinfected and should, therefore, be wrapped in disposable plastic material before each use.

FIGURE 1-4

Sandbags used for radiographic positioning.

Beanbags

Pads filled with polyester beads are commonly referred to as beanbags. They are similar in construction to sandbags and available in a variety of sizes. Beanbags with vacuum hose connections are also available. This allows the handler to mold the bag around the patient by removing the air from the bag while it is in position on the patient. Beanbags are usually composed of vinyl or similar materials.

Foam Pads and Wedges

Foam pads and wedges are available in a wide variety of sizes and shapes (Figure 1-5). Triangular and rectangular foam blocks are most common. Some foam wedges and pads have heavy vinyl covers that are washable. Plain foam pads and wedges cannot be easily disinfected and must be covered with disposable materials before use. Foam positioning aids are usually radiolucent, although some that are covered with heavy fabrics may leave density shadows on the finished radiograph.

Troughs

U- and V-shaped troughs are commonly used positioning aids. They are available in a variety of widths and lengths. Troughs are designed to maintain a patient in dorsal recumbency. They are commonly composed of clear plastic (Figure 1-6) or may be vinyl-covered,

FIGURE 1-5

Foam wedges.

FIGURE 1-6

Clear plastic V-trough.

trough-shaped foam (Figure 1-7). Plastic troughs are radiolucent. Head troughs are also available that are U-shaped and contain acrylic rods that can be used to maintain the position of the skull. When radiographing the thorax or abdomen using a positioning trough, the length of the trough must be sufficient to allow the entire area of interest to remain fully within the trough. If using the trough to position an animal for other radiographic studies, such as the pelvis, the trough must be fully outside the collimated area of the film. If the trough is not a sufficient length or is positioned improperly, the finished radiograph may contain density artifacts representing the edges of the trough.

FIGURE 1-7

Vinyl-covered V-trough.

Other Positioning Aids

Rope, roll gauze, adhesive tape, and plastic tubing make inexpensive and effective positioning aids. Adhesive tape can serve many functions as a positioning aid. It can be used to extend a limb or widen the space between the digits to increase visibility of the structures. It can also be used to rotate limbs and hold them in position to obtain oblique views, or maintain bones and joints perpendicular to the x-ray beam. Rope and gauze can be looped around a limb and used to extend the limb. The end of the rope can be tied to the table or can be held by the handler. Extending the limb by holding the end of the rope or gauze increases the distance of the handler to the primary beam, thereby reducing exposure to ionizing radiation. Compression bands and paddles, wooden spoons, and Velcro can also be used to assist in immobilizing patients. A plastic-mouth speculum may be useful for positioning of some dental and skull radiographs.

Acrylic tubes are routinely used for immobilization of exotic animals for radiography. Although chemical restraint is generally required for most exotic animal patients, paper bags and pillowcases may be used to contain some exotic animal species. Detailed information on positioning aids used in exotic animals is presented in Chapter 10. A large number of specialized radiographic positioning aids are also commercially available for use in small animals.

PROCEDURES FOR RADIOGRAPHIC POSITIONING

Specific protocols for positioning of animals for radiographic evaluation vary depending on the anatomical area of interest and the species. For nearly all radiographic studies, two views are needed, taken at right angles to each other. Patients are usually positioned with the area of interest as close to the x-ray cassette as possible. This decreases magnification and increases detail. In some cases, magnification is desirable, and the radiograph exposure will be made with the area of interest elevated off the x-ray cassette so that it is closer to the x-ray tube. This is commonly performed when obtaining radiographs of exotic animals. Except for oblique views and some dental radiographs, the area of interest should remain perpendicular to the x-ray tube to minimize distortion of the area of interest on the resulting radiograph.

The patient should be placed on the x-ray table so that the thickest part of the area of interest is placed toward the cathode end of the x-ray tube. This takes advantage of the heel effect, which refers to the greater intensity of x-rays that originate from the cathode end of the tube, and results in more even film density on the finished radiograph.

Centering and Collimation

The specific anatomical structures that must be included on the finished radiograph are based on surface landmarks. These are fixed areas on the patient's body that can be seen or palpated. For example, the last rib, the angle of the mandible, and the scapula are useful palpable landmarks.

Animals should always be positioned so that the anatomical area of interest for the radiographic study is at the center of the x-ray film. For example, when survey radiographs of the abdomen are taken, the primary x-ray beam is centered on the midline. However, if the anatomical area of interest is a canine patient's spleen, the primary beam would be centered lateral to the midline so that the center of the spleen is located in the center of the finished radiograph.

Radiographs should include a large enough area to allow identification of the structures. For example, radiographs of long bones must include the joints proximal and distal to the bone, whereas radiographs of joints must include 1/3 of the bones proximal and distal to the joint.

The cassette size chosen must be slightly larger than needed to accommodate the needed view. The collimator is then used to restrict the size of the beam (Figure 1-8). This results in a reduction of scatter radiation, thereby reducing exposure of the handler to the primary beam as well as improving the overall quality of the image. Properly collimated films will have a clear, unexposed area on all four sides of the finished radiograph. When the x-ray beam is centered correctly and the correct size cassette is used, it is usually only necessary to verify one surface landmark on the Cr/Cd aspect of the patient and one on the D/V aspect.

In some cases, it is desirable to use an x-ray cassette twice as large as needed and expose one view on each side of the film. This is commonly referred to as "splitting the plate." It is accomplished using a lead shield across half of the x-ray cassette to prevent exposure of the film when taking the first view (Figure 1-9). The lead shield is then moved to the other half of the cassette, and the second view is exposed on the other half of the film. It is important that the patient is oriented in the same direction each time so that the finished radiograph has the two views facing the same direction.

Measurement

A caliper is used to measure the patient so that the correct settings can be chosen on the x-ray machine (Figure 1-10). The measurement is obtained over the thickest part of the body in the area to be radiographed. Where there is a significant difference in size between the cranial and caudal surface landmarks, it may be necessary to use two cassettes to obtain the needed views. In that case, separate radiographs are taken of the cranial and caudal portions of the area of interest. This is most common in large, barrel-chested dogs requiring abdominal or thoracic radiographic studies. When using a V-trough that will be positioned within the collimated area, the trough is included in the total measurement taken with the caliper.

FIGURE 1-8

Collimator.

FIGURE 1-9

Split plate.

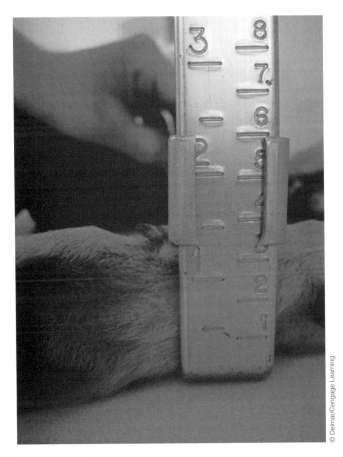

FIGURE 1-10

Caliper used to obtain body measurement.

FIGURE 1-11

Radiographic label imprinter.

FIGURE 1-12

Lead radiograph label tape on density filters.

Labeling

There are several acceptable methods for labeling radiographs. Patient information may be written on lead tape placed on the film before exposure or may be imprinted on the film in the darkroom. When an imprinter is used (Figure 1-11), the x-ray cassette must contain a lead blocker to shield a portion of the film from exposure. Printed cards are used that are prepared with the patient information. The imprinter is used in the darkroom to flash white light through the previously unexposed area of the cassette to transfer information from the printed card to the x-ray film. Lead tape is used with density filters that are matched to the exposure settings for that radiographic study. In general, radiographic studies that utilize the grid with the cassette under the tabletop require a green density filter, whereas exposures made with the cassette on the tabletop utilize the white density filter (Figure 1-12). Digital systems usually utilize computer software to imbed the patient information on the finished radiograph. Minimal required information includes the date the radiograph was exposed, the name

of the veterinarian or clinic, and the patient and client name. In addition to the patient information, the identification label may also designate the direction of entry and exit of the primary beam. Radiographic projections of limbs may also be designated as forelimb or hindlimb on the film label. The identification label must be placed so that it does not overlap any anatomical area of interest.

Directional markers must also be included on all radiographs. These markers are added before making the exposure. Directional markers can be composed of a lead shield cut to an *R* or *L* shape, or can be obtained commercially. A common type of directional marker is composed of metal with an *R* or *L* cut out. Markers are used to designate the position of the patient on its right or left side as well as the limb being radiographed. Markers on craniocaudal or caudocranial projections are placed on the lateral aspect of the limb. For lateral projections of limbs, the left or right marker is placed on the cranial aspect. For dorsoventral or ventrodorsal projections, the marker is used to indicate the right or left side of the patient. Multiple types of markers are

(a)

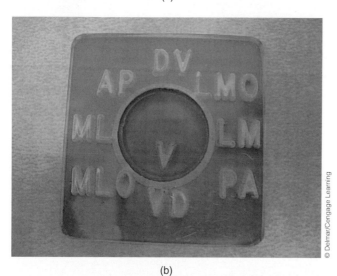

(b)

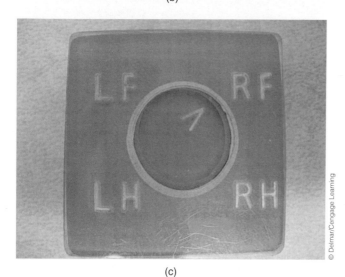

(c)

FIGURE 1-13a—c

Directional markers.

available that can be used to designate detailed information on the patient position (Figures 1-13a—c).

Some radiographic contrast studies require exposure of sequential radiographs. These films must also be designated with a time marker (Figure 1-14). This usually refers to the elapsed time since the radiographic study was started or can indicate the number of the radiograph in the series. The designation can be made on the lead tape identification label. Timer markers are also available that contain a clockface with rotating dials to indicate the elapsed time or the time the radiograph was exposed. Gravity markers that can be used to designate that the patient is standing are also available.

FIGURE 1-14

Elapsed time marker.

CHAPTER 2

THORACIC RADIOGRAPHS

OVERVIEW

Thoracic radiographs are primarily utilized for evaluation of the soft tissues of the thoracic cavity (i.e., lungs, heart). Thoracic radiographs are usually exposed at peak inspiration. In patients with suspected pneumothorax, exposures are usually made during the expiratory pause. The most commonly used positions are the right or left lateral recumbency and ventrodorsal (VD). If VD, dorsoventral (DV), and right and left lateral views are needed, the DV and VD exposures should be performed first to prevent positional collapse of the lungs.

For all thoracic projections, the forelimbs must be extended cranially to avoid overlap of the shoulder muscles on the thoracic structures. For the DV and VD projections, the sternum appears superimposed on the thoracic vertebrae. In properly positioned lateral projections, the costochondral junctions of the ribs and the angles of the thoracic vertebrae are even and superimposed in some areas. A horizontal beam may occasionally be used to identify the presence of air or fluid within the thoracic cavity. Lateral, DV, and VD views may be exposed with the horizontal beam.

The following pages illustrate the proper positioning and technique for thoracic radiographs.

Lateral Projection of the Thorax

Positioning:

- Right lateral recumbency is preferred.
- Forelimbs are extended cranially; hindlimbs caudally.
- Place a foam pad under the sternum to avoid rotation and to maintain horizontal alignment of the sternum and spine.
- Neck is in natural position.

Centering:

- Caudal border of scapula.

Collimation:

- Entire rib cage within collimated area.
- Cranial border: thoracic inlet.
- Dorsal border: spinous processes of spinal column.
- Ventral border: xiphoid.

Labeling:

- R/L marker within collimated area ensuring that marker does not obscure any anatomical structure.
- Identification label in caudal region within collimated area.

Technique:

- Measure at highest area.

Comments:

- Broad-chested animals may not require foam padding to position sternum.
- Ensure that any sandbags are placed outside the collimated area.

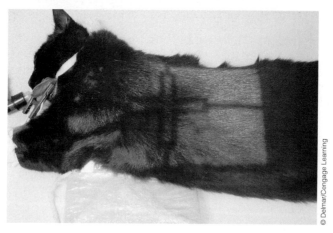

© Delmar/Cengage Learning

FIGURE 2-1

Proper positioning for lateral projection of the thorax.

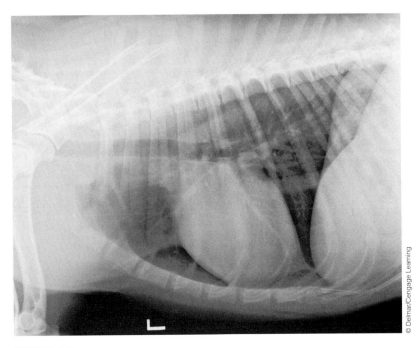

FIGURE 2-2

Lateral projection of the thorax.

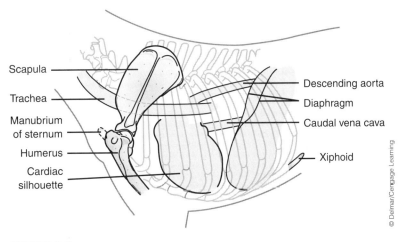

FIGURE 2-3

Anatomical features and landmarks: cardiac silhouette, caudal vena cava, trachea, descending aorta, scapula, diaphragm, humerus, manubrium of sternum, and xiphoid.

VD Projection of the Thorax

Positioning:

- Dorsal recumbency.
- Forelimbs are extended cranially with nose between forelimbs.
- Hindlimbs are extended caudally.
- Use a V-trough to superimpose the sternum and spine.

Centering:

- Caudal border of the scapula centered on midline.

Collimation:

- Cranial border: thoracic inlet.
- V-trough entirely within collimated area.
- Lateral borders: body wall.

Labeling:

- R/L marker cranial to axilla within V-trough.
- Identification label in cranial or caudal region within collimated area.

Technique:

- Measure at highest point (usually the last rib).

Comments:

- Foam pads may be needed to maintain alignment of the sternum and spine, and to avoid rotation.
- This position may also be used with horizontal beam x-ray and is referred to as the lateral decubitus view (Figure 2-5). The patient is placed in lateral recumbency on top of a foam pad to elevate the patient off the tabletop. The beam is then directed ventrodorsally.

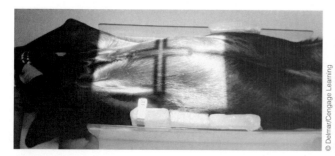

© Delmar/Cengage Learning

FIGURE 2-4

Proper positioning for VD projection of the thorax.

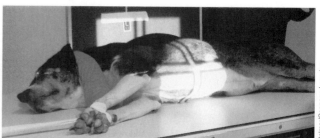

© Delmar/Cengage Learning

FIGURE 2-5

Lateral decubitus view.

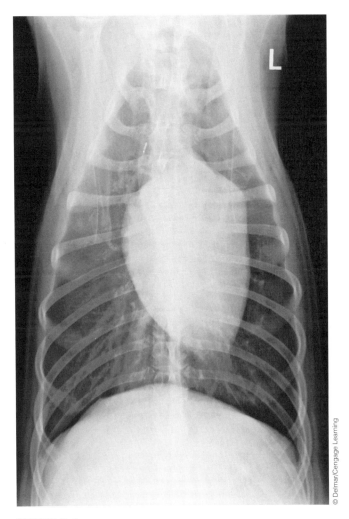

FIGURE 2-6

VD projection of the thorax.

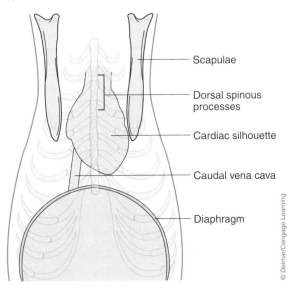

Scapulae

Dorsal spinous processes

Cardiac silhouette

Caudal vena cava

Diaphragm

FIGURE 2-7

Anatomical features and landmarks: scapulae, cardiac silhouette, caudal vena cava, diaphragm, and dorsal spinous processes.

DV Projection of the Thorax

Positioning:

- Sternal recumbency.
- Forelimbs are extended slightly cranial with carpus at level of ears.
- Hindlimbs are in natural flexed position.
- Superimpose sternum and spine.

Centering:

- Caudal border of scapula.

Collimation:

- Cranial border-thoracic inlet.
- Lateral borders-body wall.

Labeling:

- R/L marker caudal to axilla within collimated area.
- Identification label in caudal region within collimated area.

Technique:

- Measure at highest point (usually the last rib).

Comments:

- Maintain alignment of sternum and spine.
- This position may also be used for horizontal x-ray beam.

FIGURE 2-8

Proper positioning for DV projection of the thorax.

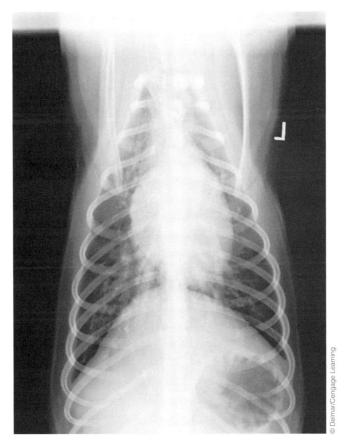

FIGURE 2-9

DV projection of the thorax.

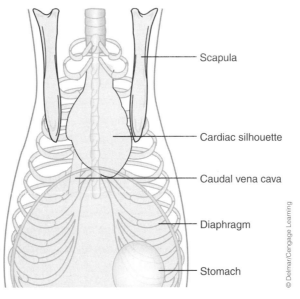

FIGURE 2-10

Anatomical features and landmarks: scapula, cardiac silhouette, stomach, diaphragm, caudal vena cava, dorsal spinous processes, and rib.

Standing Lateral Projection of the Thorax Using the Horizontal Beam

Positioning:

- Right lateral preferred.
- Natural standing position.

Centering:

- Caudal border of scapula.

Collimation:

- Entire rib cage within collimated area.
- Cranial border: thoracic inlet.
- Dorsal border: spinous processes of spinal column.
- Ventral border: sternum.

Labeling:

- R/L marker cranial to axilla within collimated area.
- Identification label in caudal region within collimated area.

Technique:

- Measure at highest area.

Comments:

- Gravitational markers, such as the Mitchell marker, should be used.
- Shoulder musculature is superimposed over the cranial thorax.
- The horizontal beam can also be used for a recumbent lateral view. The patient is positioned in sternal recumbency on top of a foam pad with the forelimbs gently extended forward and the hindlimbs in a natural crouched position (Figure 2-12).

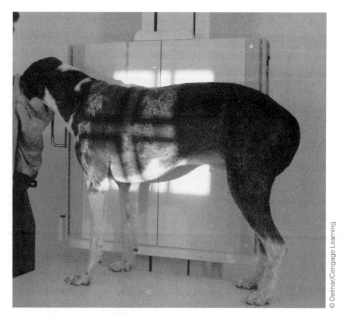

FIGURE 2-11

Proper positioning for standing lateral projection with horizontal beam of the thorax.

FIGURE 2-12

Lateral recumbent view.

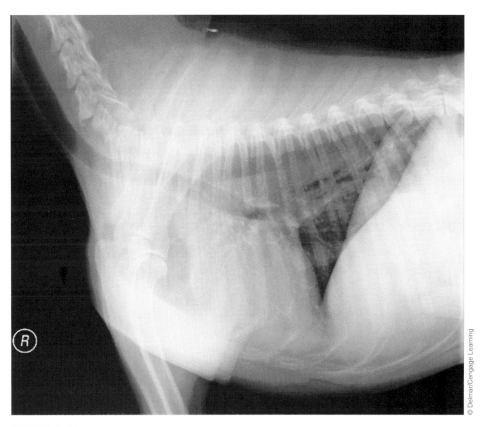

FIGURE 2-13

Standing lateral projection with horizontal beam of the thorax.

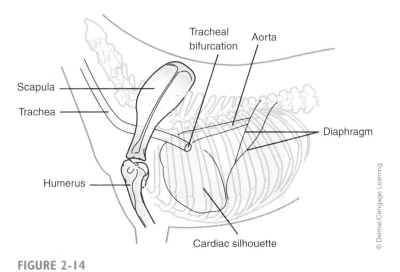

FIGURE 2-14

Anatomical features and landmarks: trachea, aorta, tracheal bifurcation, cardiac silhouette, scapula, humerus, and diaphragm.

CHAPTER 3

ABDOMINAL RADIOGRAPHS

OVERVIEW

Abdominal radiographs are primarily utilized for evaluation of the soft tissues of the abdomen (kidneys, bladder, liver, intestinal tract). Abdominal radiographs are exposed after full exhalation and before initiation of inspiration (expiratory pause). Lateral and ventrodorsal (VD) views are commonly performed.

For large dogs, it may be necessary to take cranial and caudal views of the abdomen if large cassettes do not allow the entire abdomen to be exposed on one film. Cranial views usually require modification of the exposure settings to avoid overexposure of the film.

Some radiographic studies must be performed after the patient has been fasted for 12 hours unless medical conditions contraindicate fasting. If necessary, a cathartic or enema may be given 3–4 hours before radiography to clear the intestinal tract of fecal matter. This will enhance visualization of structures within the abdominal cavity.

The following pages illustrate the proper positioning and technique for abdominal radiographs.

Lateral Projection of the Abdomen

Positioning:

- Right lateral recumbency.
- Forelimbs are extended cranially; hindlimbs extended caudally.
- Use foam pads to maintain horizontal alignment of sternum.
- Use foam pads between stifles to maintain alignment.

Centering:

- Slightly caudal to last rib.

Collimation:

- Cranial border: halfway between the caudal border of scapula and the xiphoid.
- Dorsal border: spinous processes of vertebral column.
- Ventral border: sternum.

Labeling

- R/L marker within collimated area in inguinal region.
- Identification label in caudal region within collimated area.

Technique:

- Measure at highest point (usually the last rib).

Comments:

- Collimated area should not extend beyond dorsal spinous processes, and caudal border must include the cranial aspect of the greater trochanter. Should the patient be too large to include both cranial and caudal landmarks, two radiographs must be taken so that both the cranial and caudal abdomen can be evaluated.
- Extension of hindlimbs is crucial to avoid superimposing of abdominal muscles, but hyperextension must be avoided because this may reduce visibility of abdominal organs.
- The standing or recumbent lateral views may also be used (see Figures 2-11 and 2-12 for patient positioning).

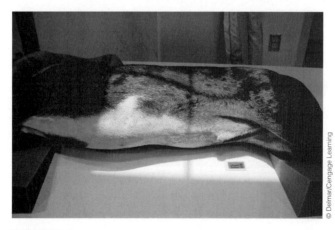

© Delmar/Cengage Learning

FIGURE 3-1

Proper positioning for lateral projection of the abdomen.

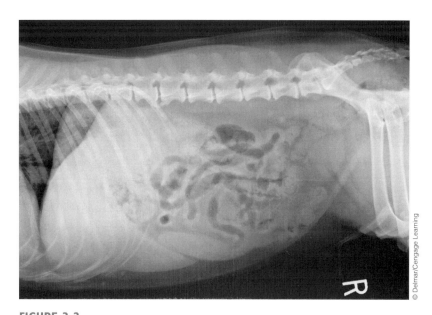

FIGURE 3-2

Lateral projection of the abdomen.

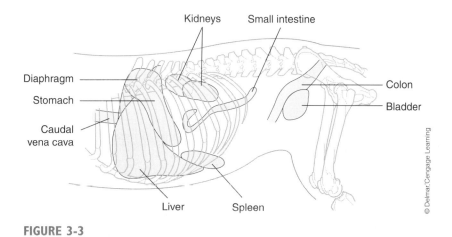

FIGURE 3-3

Anatomical features and landmarks: liver, spleen, stomach, kidneys, colon, small intestine, bladder, diaphragm, and caudal vena cava.

VD Projection of the Abdomen

Positioning:

- Dorsal recumbency.
- Forelimbs extended cranially with nose between forelimbs.
- Hindlimbs extended caudally.
- Use a V-trough to superimpose the sternum and spine.

Centering:

- Medial aspect of last rib centered on sternum.

Collimation:

- Cranial border: on midline halfway between the caudal border of scapula and the xiphoid.
- Lateral borders: abdominal wall within V-trough.

Labeling:

- R/L marker in V-trough within collimated area.
- Identification label in caudal region within collimated area.

Technique:

- Measure at highest point (usually the last rib).

Comments:

- Foam pads may be needed on lateral aspect of body wall to avoid rotation and maintain alignment of sternum and spine.
- Edges of V-trough must be outside the collimated area.
- The VD projection with the horizontal beam may also be used (see Figure 2-5 for patient positioning).

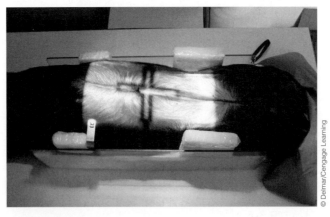

© Delmar/Cengage Learning

FIGURE 3-4

Proper positioning for VD projection of the abdomen.

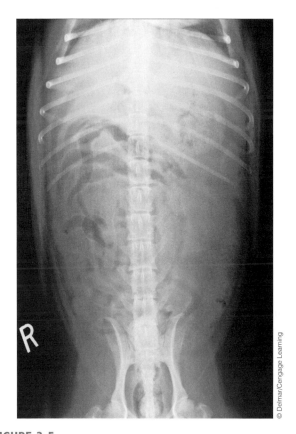

FIGURE 3-5

VD projection of the abdomen.

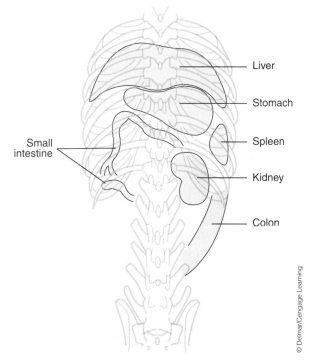

FIGURE 3-6

Anatomical features and landmarks: stomach, spleen, kidney, colon, liver, and small intestine.

CHAPTER 4

PELVIC RADIOGRAPHS

OVERVIEW

Pelvic radiographs are primarily utilized for visualization of the bones and joints that comprise the hip. In patients with suspected hip dysplasia, a variety of specialized procedures are used. The most common of these is the ventrodorsal (VD)-extended hip view utilized for certification by the Orthopedic Foundation of America. Always consult the foundation for specific application procedures and requirements.   Some practitioners use a diagnostic technique for hip dysplasia known as the PennHIP method. This procedure requires specialized equipment, and can only be performed by those who have undergone additional training and certification.

For all pelvic projections, the hip joints and sacroiliac joints are mirror images of each other. In addition to the VD-extended view, commonly performed projections of the pelvis include the VD frog leg position and lateral.

The following pages illustrate the proper positioning and technique for pelvic radiographs.

VD-Extended Hip Projection

Positioning:

- Dorsal recumbency.
- Forelimbs extended cranially and evenly with nose between forelimbs.
- Hindlimbs extended caudally and evenly into full extension.
- V-trough with foam pads on lateral aspect of body wall to superimpose sternum and spine.
- Femurs rotated medially so they are parallel to one another and the x-ray table, and the patella is centered within the patellar groove over the stifle and taped in place.
- Align tail with spine.

Centering:

- Midline between the left and right ischial tuberosity.

Collimation:

- Cranial border: caudal to the wing of ilium and distal to the patella.
- Lateral borders: lateral to the ischium.

Labeling:

- R/L marker within collimated area away from bony areas.
- Detailed permanent identification in caudal region to include American Kennel Club-registered name, number or case number, hospital or veterinarian name, and date radiograph was taken.

Technique:

- Measure thickest part of pelvis.

Comments:

- Hindlimbs must both be extended evenly so that digits on both feet are even with each other.
- A long piece of tape can be used to rotate femurs by applying tape sticky side up under the stifles, excluding the tail. Pull each end of the tape to the opposite side of the table, using the tape to pull the femurs medially. Hold the ends of the tape in place with sandbags, using the sandbag to place additional pressure on the tape.

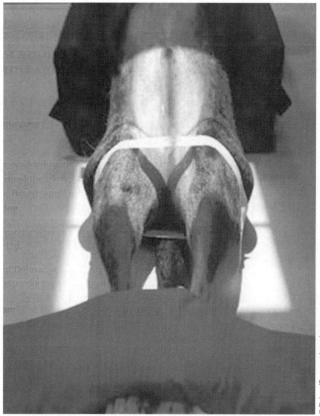

© Delmar/Cengage Learning

FIGURE 4-1

Proper positioning for VD-extended hip.

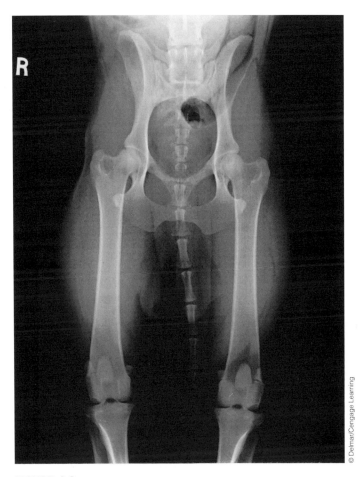

FIGURE 4-2

VD-extended hip projection.

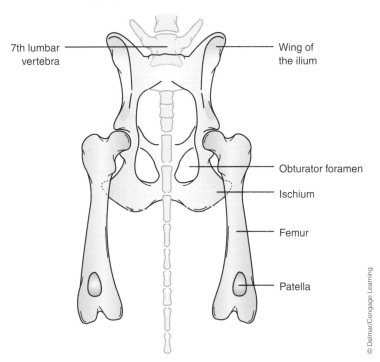

FIGURE 4-3

Anatomical features and landmarks: femur, patella, obturator foramen, wing of the ilium, ischium, and 7th lumbar vertebra.

VD Frog Leg Projection

Positioning:

- Dorsal recumbency.
- Forelimbs are extended cranially.
- Hindlimbs are in natural flexed position; in most normal patients, the femurs naturally assume an angle of approximately 45 degrees to the spine. In some large dogs, the femurs may naturally assume a 90-degree angle to the spine.
- Use a V-trough with foam pads on lateral aspect of body wall to superimpose sternum and spine.

Centering:

- Midline between the left and right ischial tuberosity.

Collimation:

- Cranial border: cranial to the wing of ilium to caudal border of ischium.
- Lateral border: to include proximal third of femur.

Labeling:

- R/L marker within collimated area away from bony areas.
- Identification label in caudal region.

Technique:

- Measure thickest part of pelvis.

Comments:

- Sandbags placed outside of collimated area on tarsus can be used to maintain symmetry.

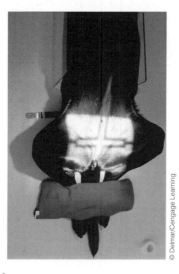

© Delmar/Cengage Learning

FIGURE 4-4

Proper positioning for VD frog leg projection.

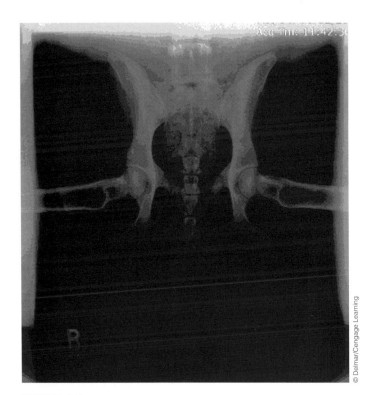

FIGURE 4-5

VD frog leg projection.

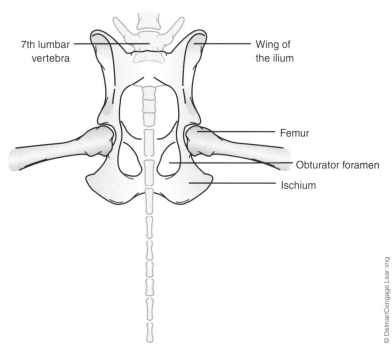

FIGURE 4-6

Anatomical features and landmarks: femur, obturator foramen, ischium, wing of the ilium, and 7th lumbar vertebra.

Lateral Projection of the Pelvis

Positioning:

- Right or left lateral recumbency (side of interest closest to the cassette).
- Foam wedge placed between hindlimbs to superimpose both sides of pelvis.
- Bottom leg extended cranially, top leg extended caudally (scissor position).

Centering:

- Greater trochanter of femur.

Collimation:

- Cranial edge of ilium to caudal border of ischium.
- Dorsal border: dorsal to the wing of the ilium.

Labeling:

- R/L marker to indicate which limb is closest to the cassette.
- Place identification label in right cranial region or left caudal region within collimated area to avoid overlap of bone.

Technique:

- Measure highest area at level of trochanter.

Comments:

- Limb furthest from cassette will be magnified.

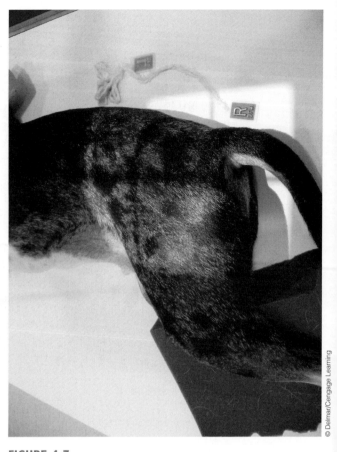

FIGURE 4-7

Proper positioning for lateral pelvis projection.

© Delmar/Cengage Learning

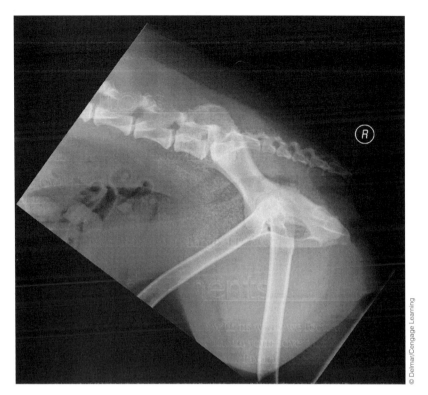

FIGURE 4-8

Lateral pelvis projection.

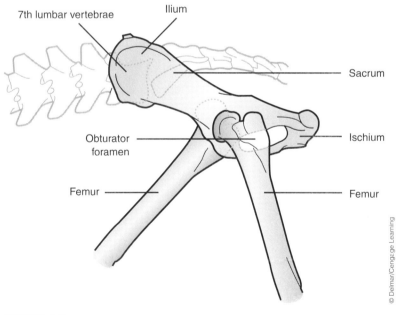

FIGURE 4-9

Anatomical features and landmarks: femur, ilium, sacrum, 7th lumbar vertebrae, obturator foramen, and ischium.

CHAPTER 5

FORELIMB RADIOGRAPHS

OVERVIEW

Radiographic projections of the limbs of the thoracic girdle are often performed to detect fractures. Careful positioning is needed to maintain the limb in a parallel plane against the x-ray cassette to avoid magnification and distortion of the image. The x-ray cassette is normally placed on the tabletop rather than under the table due to the relatively small measurement of dog and cat limbs. Collimation includes joints above and below a bone for images of long bones. Radiographic projections of joints generally include approximately one-third the bones proximal and distal to the joint. The beam is restricted to just the width needed to include all the necessary structures. This reduces scatter radiation and results in a higher quality image. When patients are exhibiting signs of pain, a horizontal beam may be used to minimize the need to manipulate the limb. Supporting the limb on a foam pad and placing the x-ray cassette perpendicular to the tabletop accomplish the horizontal beam projection.

Radiographic evaluation of the forelimbs includes lateral and caudocranial (CdCr) images of the scapula, humerus, shoulder, elbow joint, radius, and ulna. Dorsopalmar and lateral views are usually taken of the carpus, metacarpus, and phalanges. Oblique views are often needed for the carpus, and flexed views of the elbow and carpus are routinely performed.

CdCr Projection of the Shoulder CdCr

Positioning:

- Dorsal recumbency in a V-trough with affected limb down.
- Tape and extend both forelimbs cranially.
- Head is pushed laterally away from the dependent limb to avoid having the cervical spine superimposed over joint.

Centering:

- Palpate proximal head of the humerus and the glenoid of the scapula. Center beam at the indentation palpated.

Collimation:

- Include the distal third of the scapula and proximal third of the humerus.

Labeling:

- Lateral to the joint.

Technique:

- Measure at the shoulder joint.

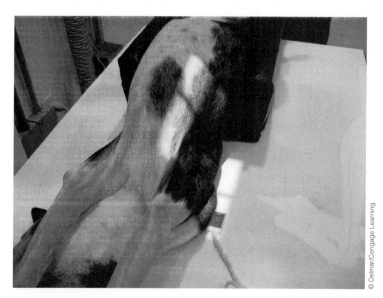

FIGURE 5-1

Proper positioning for CdCr projection of the shoulder CdCr.

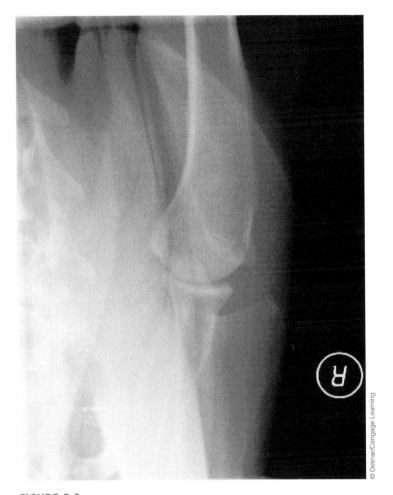

FIGURE 5-2

CdCr projection of the shoulder CdCr.

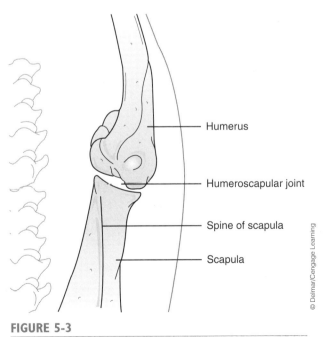

Humerus

Humeroscapular joint

Spine of scapula

Scapula

FIGURE 5-3

Anatomical features and landmarks: scapula, spine of scapula, humerus, and humeroscapular joint.

Lateral Projection of the Shoulder

Positioning:

- Lateral recumbency with affected limb down.
- Affected limb is extended cranially.
- Opposite limb is pulled caudally to eliminate superimposition.
- Head is pushed dorsally to avoid trachea overlying joint.
- Sandbags can be used to keep head in dorsal position.

Centering:

- Palpate proximal head of the humerus and the glenoid of the scapula.
- Center beam approximately 1 inch (2.5 cm) caudal from the front of the limb for a small dog and approximately 2 inches caudal (5 cm) for a large dog.

Collimation:

- Include the distal third of the scapula and proximal third of the humerus, excluding the upper leg that is pulled caudally.

Labeling:

- Cranial to the joint.

Technique:

- Measure at the shoulder joint, being careful not to include the upper leg that is pulled caudally.

FIGURE 5-4

Proper positioning for lateral projection of the shoulder.

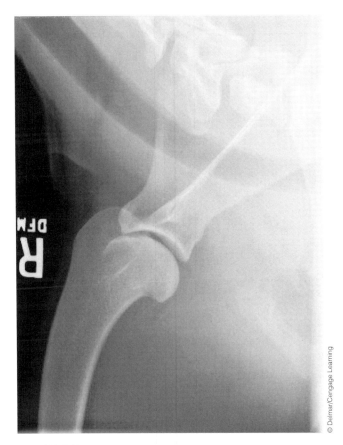

FIGURE 5-5

Lateral projection of the shoulder.

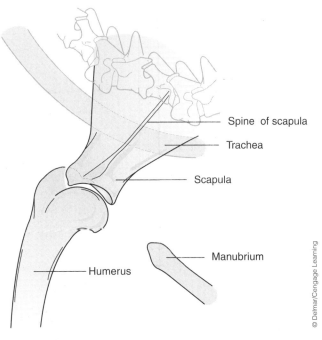

FIGURE 5-6

Anatomical features and landmarks: scapula, spine of scapula, humerus, trachea, and manubrium.

CdCr Projection of the Scapula CdCr

Positioning:

- Dorsal recumbency.
- V-trough to hold body straight with cranial half of thorax outside the trough.
- Legs extended forward individually.
- Head straight with spine.
- Back legs extended caudally to stabilize.

Centering:

- Center of the scapula.

Collimation:

- Lateral to the body and medial to the spine.
- Include the shoulder joint and the caudal border of scapula.

Labeling:

- Lateral to the scapula.

Technique:

- Measure at cranial border of the scapula.

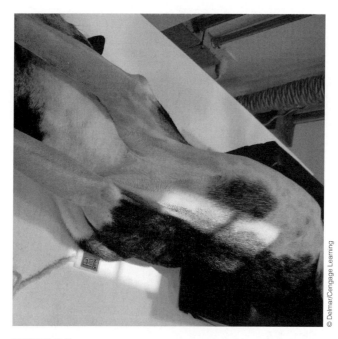

FIGURE 5-7

Proper positioning for CdCr projection of the scapula CdCr.

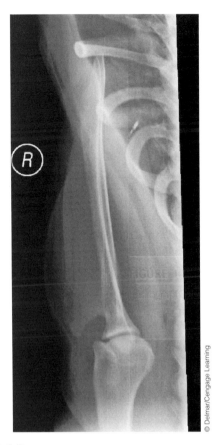

FIGURE 5-8

CdCr projection of the scapula CdCr.

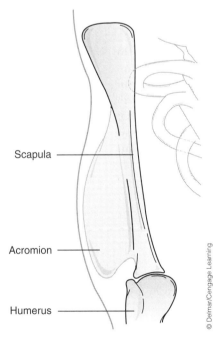

FIGURE 5-9

Anatomical features and landmarks: humerus, acromion, and scapula.

Lateral Projection of the Scapula

Positioning:

- Patient is in lateral recumbency with affected scapula up.
- Unaffected leg is extended forward.
- Affected limb is pushed up dorsally and stabilized with sandbag to push the scapula above the thoracic spine.
- Skull and neck are pushed downward and stabilized with sandbag, if necessary.

Centering:

- Center of the scapula.

Collimation:

- Proximal to the shoulder joint to the caudal edge of the scapula.

Labeling:

- Place dorsally and with marker to indicate the affected limb.

Technique:

- Measure from the dorsal side from table to height of the scapula.

Comments:

- Scapula will be superimposed over the dorsal spinous processes. Lateral scapula view can also be taken with the affected side down.
- Push the affected limb dorsally and secure with sandbag.
- Pull the head ventrally to avoid superimposition of the cervical spine.
- Extend the unaffected limb straight out and away from the body with the limb parallel to the table-top, and then move the limb cranial as far as possible to avoid superimposition.
- The scapula is a bone density within air of the thorax. The thoracic exposure technique is not sufficient and bone technique will be too dark. The abdominal technique chart should be used when calculating the exposure settings.

© Delmar/Cengage Learning

FIGURE 5-10

Proper positioning for lateral projection of the scapula.

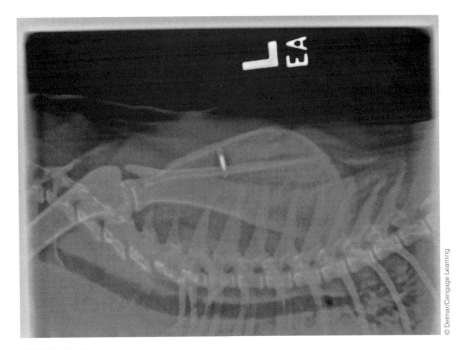

FIGURE 5-11

Lateral projection of the scapula.

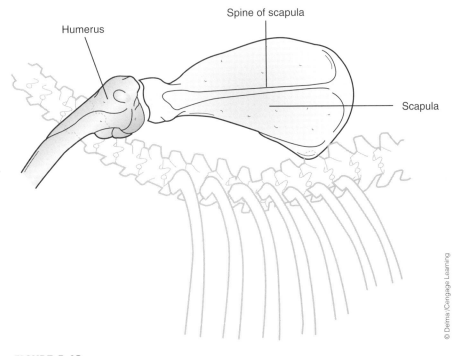

FIGURE 5-12

Anatomical features and landmarks: humerus, scapula, and spine of scapula.

CdCr Projection of the Humerus

Positioning:

- Dorsal recumbency.
- Front legs are extended forward individually.
- If needed, tape legs together at the elbows to align and straighten the humerus.

Centering:

- Midshaft of humerus

Collimation:

- Proximal to the shoulder joint to distal to the elbow joint.

Labeling:

- Place lateral to the affected limb.

Technique:

- Measure from table to midshaft humerus.

Comments:

- Patients with severe degenerative joint disease may not be able to tolerate this position.
- The alternative is to pull the humerus downward, and image cranial to caudal (Figures 5-14, 5-17, and 5-18). A separate caudal to cranial shoulder projection may be needed.

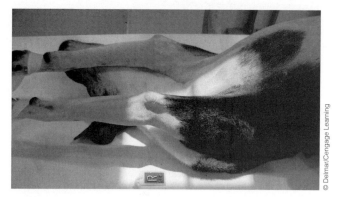

FIGURE 5-13

Proper positioning for CdCr projection of the humerus.

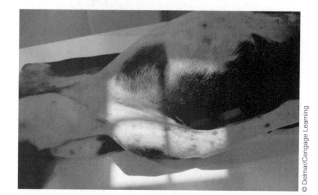

FIGURE 5-14

Alternate positioning for CdCr projection of the humerus.

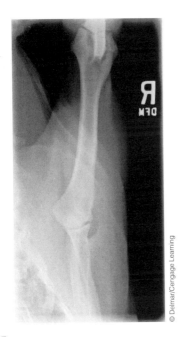

FIGURE 5-15

CdCr projection of the humerus.

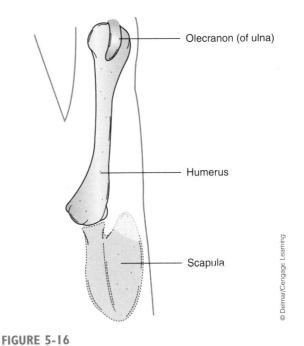

FIGURE 5-16

Anatomical features and landmarks: scapula, humerus, and olecranon.

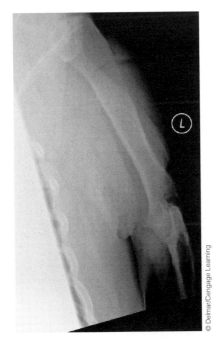

FIGURE 5-17

CdCr projection of the humerus with alternate positioning technique.

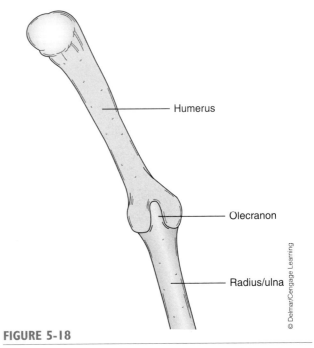

FIGURE 5-18

Anatomical features and landmarks: humerus, radius/ulna, and olecranon.

Lateral Projection of the Humerus

Positioning:

- Patient in lateral recumbency with the affected limb down.
- The affected limb is extended down and forward.
- The skull and neck are moved dorsally and stabilized with sandbags if necessary.
- The unaffected top limb is extended up caudally to pull the shoulder off the affected limb.

Centering:

- Midshaft of the humerus.

Collimation:

- Proximal to the shoulder joint to distal to the elbow joint.

Labeling:

- Place cranially.

Technique:

- Measure midshaft erring on the proximal side.

Comments:

- Larger dogs may need two views due to the thickness difference between the elbow and shoulder. Take separate measurements for each of the two views at the thickest area.

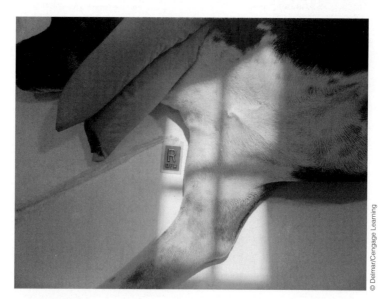

© Delmar/Cengage Learning

FIGURE 5-19

Proper positioning for lateral projection of the humerus.

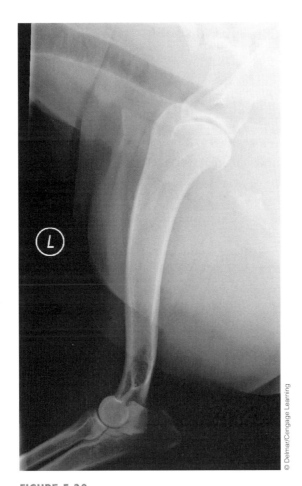

© Delmar/Cengage Learning

FIGURE 5-20

Lateral projection of the humerus.

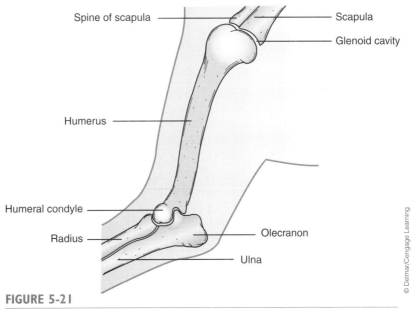

Spine of scapula ——————   —————— Scapula

                                                     —————— Glenoid cavity

Humerus ——————

Humeral condyle ——————

Radius ——————   —————— Olecranon

                    —————— Ulna

© Delmar/Cengage Learning

FIGURE 5-21

Anatomical features and landmarks: ulna, radius, humerus, scapula, glenoid cavity, olecranon, humeral condyle, and spine of scapula.

CrCd Projection of the Elbow CrCd

Positioning:

- Patient is in ventral (sternal) recumbency.
- Both front legs are extended forward individually.
- Head is extended laterally to the opposite side of the affected limb and stabilized with tape or sandbag.
- V-trough can help stabilize caudal half of body.
- Back legs may be extended caudally to assist in keeping spine straight.

Centering:

- Palpate and center on the humeral condyles.

Collimation:

- From the distal third of the humerus to the proximal third of the radius and ulna.

Labeling:

- Placed laterally.

Technique:

- Measure the thickest part at the center of the joint.

Comments:

- The horizontal beam projection is accomplished by placing the limb on sponges to have the limb extending straight out from the body (Figures 5-25 and 5-26). Center the beam on the joint as would be done in the ventral position. Tape or stabilize cassette with a sandbag.
- An alternate image includes the CrCd pronated position used when evaluating for presence of OCD lesions on the medial condyle, which requires a 10–15 degree oblique view.

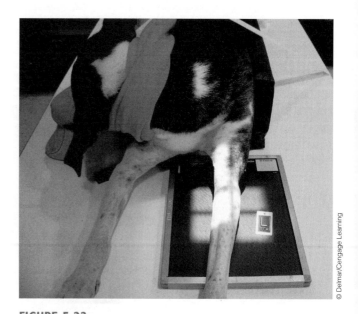

FIGURE 5-22

Proper positioning for CrCd projection of the elbow.

© Delmar/Cengage Learning

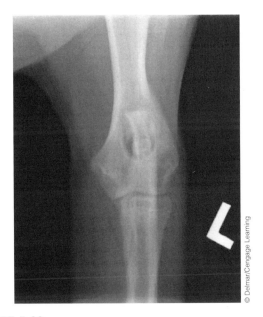

FIGURE 5-23

CrCd projection of the elbow.

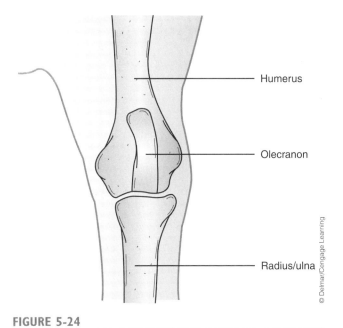

Humerus

Olecranon

Radius/ulna

FIGURE 5-24

Anatomical features and landmarks: humerus, radius/ulna, and olecranon.

FIGURE 5-25

Proper positioning for CrCd projection of the elbow with the horizontal beam.

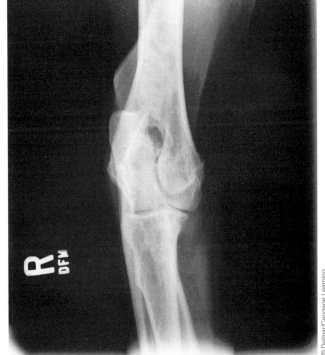

FIGURE 5-26

CrCd projection of the elbow with the horizontal beam.

Lateral Projection of the Elbow

Positioning:

- Patient is in lateral recumbency with affected limb down.
- Extend affected limb cranially.
- Pull unaffected limb caudodorsally.

Centering:

- Palpate and center on the humeral condyles.

Collimation:

- Distal third of the humerus to the cranial third of the radius and ulna.

Labeling:

- Cranial to the joint.

Technique:

- Measure the thickest part at the joint.

Comments:

- May need to place sponge under shoulder dorsally to assist with making patient lateral.
- An alternate image includes the CrCd supinated position, which is used when evaluating for presence of elbow dysplasia, which requires a 10–15-degree oblique view.

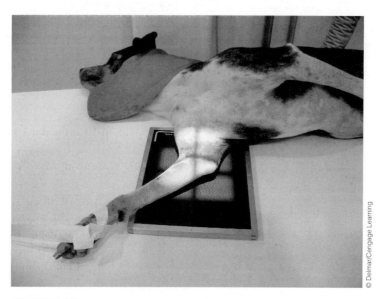

FIGURE 5-27

Proper positioning for lateral projection of the elbow.

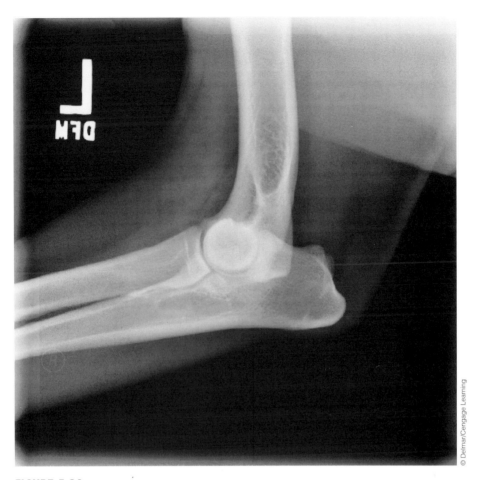

FIGURE 5-28

Lateral projection of the elbow.

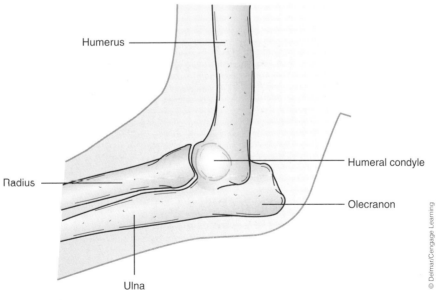

FIGURE 5-29

Anatomical features and landmarks: humerus, radius, ulna, humeral condyle, and olecranon.

Flexed Lateral Projection of the Elbow

Positioning:

- Patient is in lateral recumbency with affected limb down.
- Bend affected limb dorsally.
- Place paw under skull, and stabilize with sandbag or tape.
- Place sponge under shoulder to keep the elbow from moving medially when flexed.

Centering:

- Palpate and center on the humeral condyles.

Collimation:

- Center on joint, and include the distal third of the humerus and proximal third of radius and ulna.

Labeling:

- Place cranial to limb.

Technique:

- Measure thickest part at elbow joint. The flexed view will generally have a larger measurement than the un-flexed lateral.

Comments:

- This view is typically for younger patients when elbow dysplasia such as fragmented coronoid, ununited anconeal, and osteochondrosis are suspected. Orthopedic Foundation of America certification requires this flexed medial to lateral view.

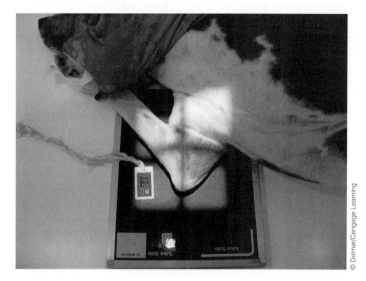

FIGURE 5-30

Proper positioning for flexed lateral projection of the elbow.

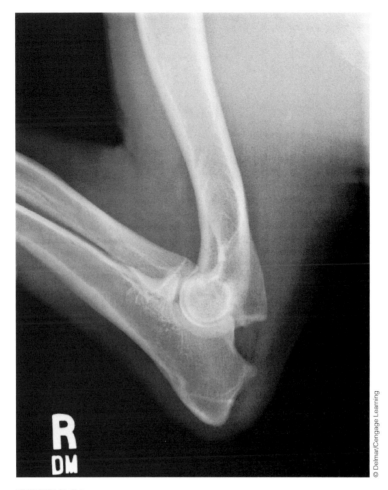

FIGURE 5-31

Flexed lateral projection of the elbow.

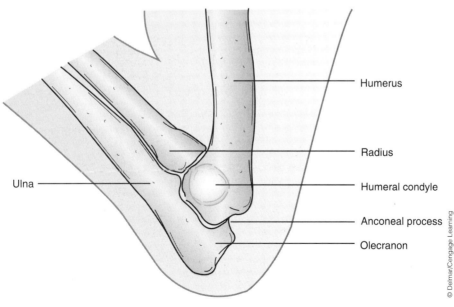

FIGURE 5-32

Anatomical features and landmarks: humerus, radius, ulna, olecranon, anconeal process, and humeral condyle.

CrCd Projection of the Radius and Ulna

Positioning:

- Patient is in ventral (sternal) recumbency.
- Front legs are extended forward individually.
- Head is extended laterally and to the opposite side of the affected limb, and stabilized with sandbag or tape.
- V-trough can help stabilize caudal half of body.
- Back legs may be extended caudally to assist in keeping spine straight.

Centering:

- Midshaft of radius and ulna.

Collimation:

- From proximal to the elbow joint and distal to the carpal joint.

Labeling:

- Place lateral to radius and ulna.

Technique:

- Measure midshaft of radius and ulna.

Comments:

- The horizontal beam projection is exposed cranial to caudal by placing the limb on sponges to have the limb extending straight out from the body (Figure 5-36). Position the beam and center on joint as would be done in the ventral position.

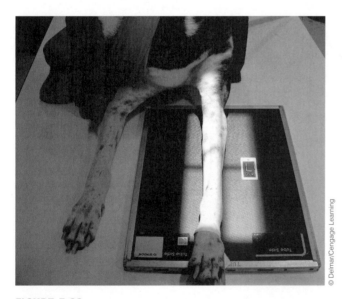

FIGURE 5-33

Proper positioning for CrCd projection of the radius and ulna.

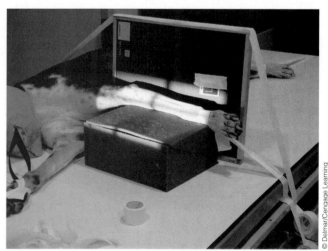

FIGURE 5-34

Proper positioning for CrCd projection of the radius and ulna with the horizontal beam.

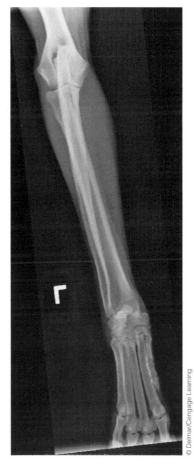

FIGURE 5-35

CrCd projection of the radius and ulna.

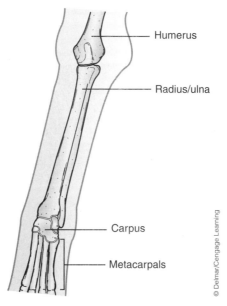

FIGURE 5-36

Anatomical features and landmarks: humerus, radius/ulna, carpus, and metacarpals.

Lateral Projection of the Radius and Ulna

Positioning:

- Patient in lateral recumbency with affected limb down.
- Unaffected limb extended caudodorsally.

Centering:

- Midshaft of the radius and ulna.

Collimation:

- From proximal to the elbow joint and distal to the carpal joint.

Labeling:

- Cranial to the radius and ulna.

Technique:

- Measure at midshaft of radius and ulna.

Comments:

- If working with a fixed (nonmovable) x-ray table, affected limb may be extended straight forward. A bend in the elbow is an acceptable position for the radius and ulna.

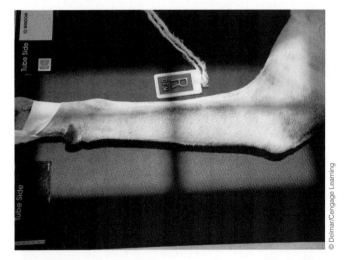

FIGURE 5-37

Proper positioning for the lateral projection of the radius and ulna.

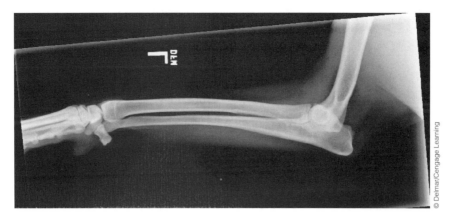

FIGURE 5-38

Lateral projection of the radius and ulna.

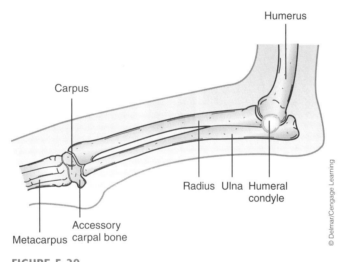

FIGURE 5-39

Anatomical features and landmarks: humerus, humeral condyle, radius, ulna, carpus, accessory carpal bone, and metacarpus.

Dorsopalmar Projection of the Carpus

Positioning:

- Patient is in ventral (sternal) recumbency.
- Front legs are extended forward individually.
- Elbow on affected limb is abducted slightly to straighten carpus.
- Head is extended laterally and to the opposite side of the affected limb, and stabilized with sandbag or tape.
- V-trough may help stabilize caudal half of body.

Centering:

- Center on carpus joint.

Collimation:

- From distal third of radius and ulna to proximal third of the metacarpals. Including all digits would also be acceptable.

Labeling:

- Lateral to carpus.

Technique:

- Measure carpal joint.

Comments:

- The horizontal beam projection may also be used. Place the limb on sponges to have the limb extending straight out from the body. Position the beam and center on joint as would be done in the ventral position.

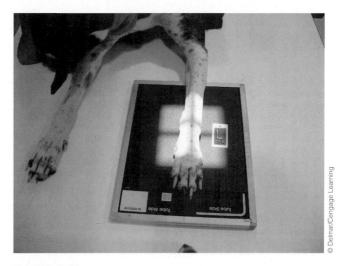

FIGURE 5-40

Proper positioning for the dorsopalmar projection of the carpus.

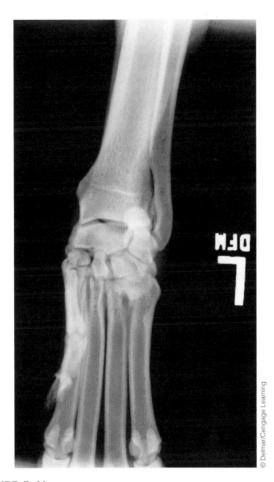

FIGURE 5-41

Dorsopalmar projection of the carpus.

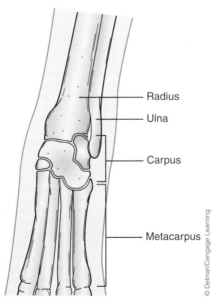

FIGURE 5-42

Anatomical features and landmarks: radius, ulna, carpus, and metacarpus.

Lateral Projection of the Carpus

Positioning:

- Patient is in lateral recumbency with affected limb down.
- Affected limb is extended down in natural position.
- Sponge is placed under elbow to make the limb even and assist with making carpus lateral.

Centering:

- Carpal joint.

Collimation:

- From distal third of radius and ulna to proximal third of the metacarpals. Including all digits would also be acceptable.

Labeling:

- Lateral to carpal joint.

Technique:

- Measure carpal joint.

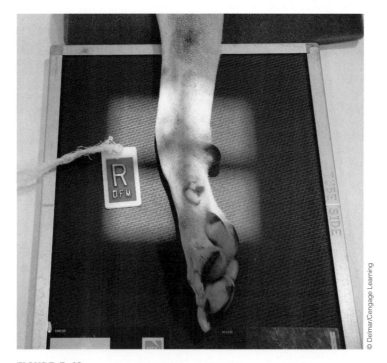

FIGURE 5-43

Proper positioning for the lateral projection of the carpus.

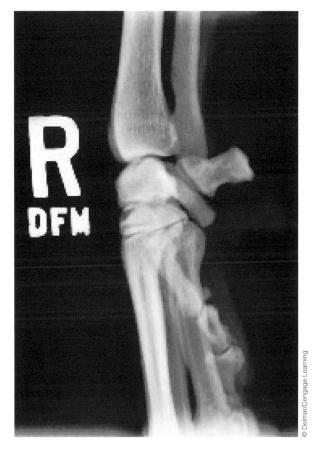

FIGURE 5-44

Lateral projection of the carpus.

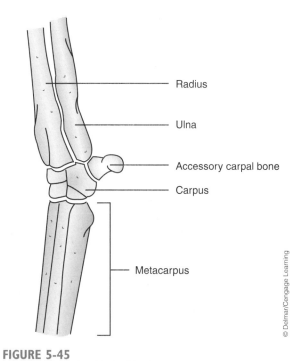

FIGURE 5-45

Anatomical features and landmarks: radius, ulna, carpus, accessory carpal bone, and metacarpus.

Flexed Lateral Projection of the Carpus

Positioning:

- Patient is in lateral recumbency with affected limb extended down naturally.
- Flex carpus by bending toes caudally toward radius and ulna.
- Keep in flexion by taping in a figure-eight pattern around metacarpals and radius and ulna.
- Place sponge under elbow to assist with maintaining lateral position.

Centering:

- Carpal joint.

Collimation:

- From distal third of radius and ulna to proximal third of metacarpals. Including all digits would also be acceptable.

Labeling:

- Cranial to flexed joint.

Technique:

- Measure thickest part of flexed joint.

Comments:

- May have to place sponge under shoulder dorsally to assist with maintaining lateral position of flexed carpus.

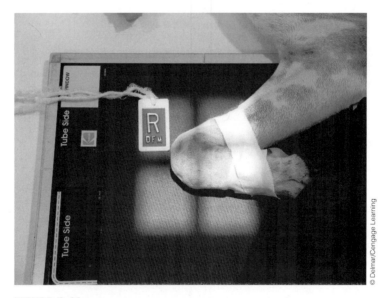

FIGURE 5-46

Proper positioning for the flexed lateral projection of the carpus.

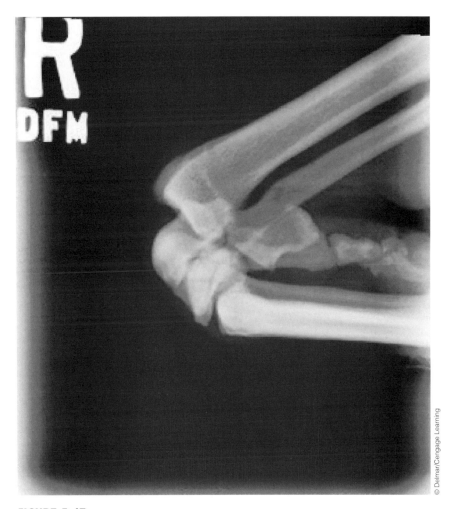

FIGURE 5-47

Flexed lateral projection of the carpus.

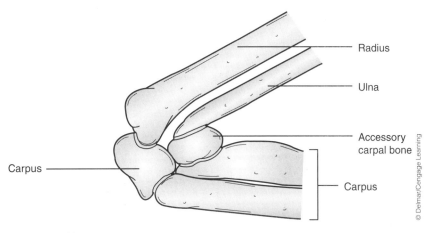

FIGURE 5-48

Anatomical features and landmarks: radius, ulna, carpus, and accessory carpal bone.

Extended Lateral Projection of the Carpus

Positioning:

- Patient is in lateral recumbency with affected limb extended down naturally.
- Extend carpus by bending toes anteriorly.
- Keep in extension by taping in a figure-eight pattern around metacarpals and radius and ulna.
- Place sponge under elbow to assist with maintaining lateral position.

Centering:

- Carpal joint.

Collimation:

- From distal third of radius and ulna to proximal third of metacarpals. Including all digits would also be acceptable.

Labeling:

- Cranial to flexed joint.

Technique:

- Measure thickest part of flexed joint.

Comments:

- May have to place sponge under shoulder dorsally to assist with maintaining lateral position of extended carpus.

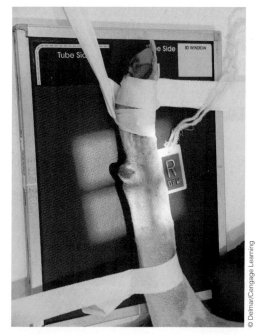

© Delmar/Cengage Learning

FIGURE 5-49

Proper positioning for the extended lateral projection of the carpus.

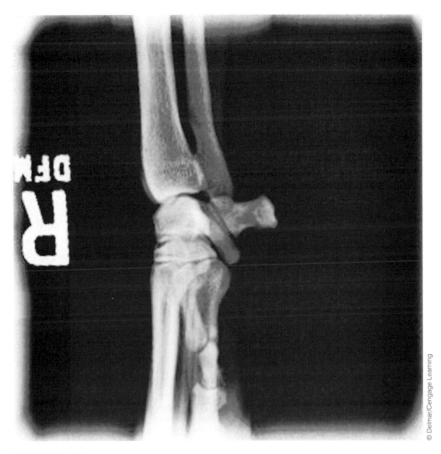

FIGURE 5-50

Extended lateral projection of the carpus.

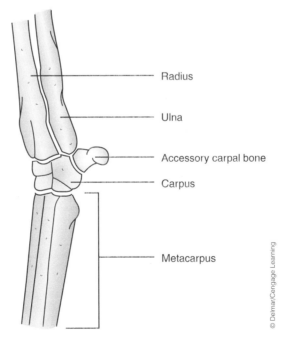

FIGURE 5-51

Anatomical features and landmarks: radius, ulna, carpus, accessory carpal bone, and metacarpus.

Lateral and Medial Oblique Projections of the Carpus

Positioning:

- Patient is in ventral (sternal) recumbency (oblique views are performed from the dorsal aspect of the joint).
- Both front legs are extended forward individually.
- Head is extended laterally to the opposite side of the affected limb and stabilized with tape or sandbag.
- V-trough can help stabilize caudal half of body.
- Hindlimbs may be extended caudally to assist in keeping spine straight.
- Lateral oblique: Pull elbow joint medially and stabilize with tape or sandbag.
- Medial oblique: Pull elbow joint laterally and stabilize with tape or sandbag.

Centering:

- Carpal joint.

Collimation:

- Distal third of radius and ulna and proximal third of metacarpals.

Labeling:

- Lateral and medial oblique markers are placed on the lateral side of both projections.

Technique:

- Measure over carpal joint.
- Measurement should remain the same for both projections.

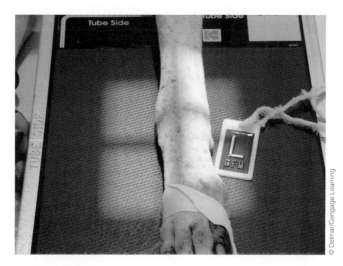

© Delmar/Cengage Learning

FIGURE 5-52

Proper positioning of the lateral and medial oblique projections of the carpus.

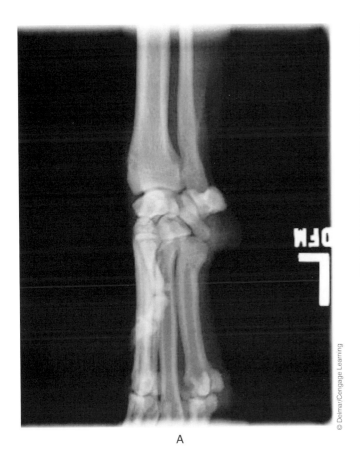
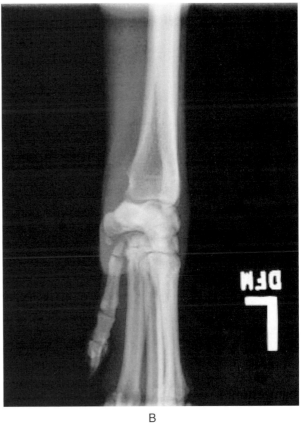

A

B

FIGURE 5-53

Lateral and medial oblique projections of the carpus.

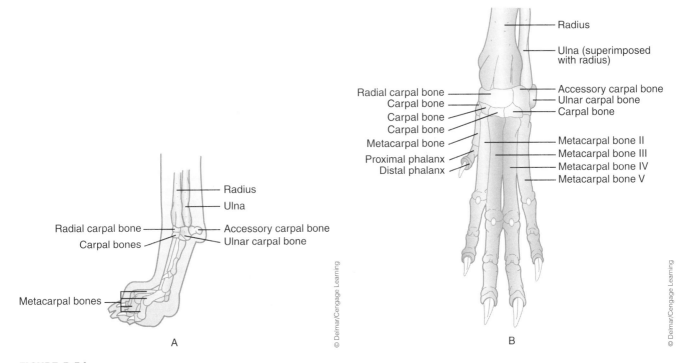

A

B

FIGURE 5-54

Anatomical features and landmarks: *A,* radius, ulna, radial carpal bone, ulnar carpal bone, accessory carpal bone, carpal bones, and metacarpal bones. *B,* radius, ulna (superimposed with radius), radial carpal bone, carpal bones, metacarpal bone, proximal phalanx, distal phalanx, accessory carpal bone, ulnar carpal bone, metacarpal bone II, metacarpal bone III, metacarpal bone IV, and metacarpal bone V.

Dorsopalmar Projection of the Metacarpals

Positioning:

- Patient is in ventral (sternal) recumbency.
- Front legs are extended forward individually.
- Elbow on affected limb is abducted slightly to straighten metacarpus.
- Head is extended laterally and to the opposite side of the affected limb, and stabilized with sandbag or tape.
- V-trough can help stabilize caudal half of body.

Centering:

- Halfway between carpus and phalanges.

Collimation:

- Distal third of radius and ulna to proximal third of the digits.

Labeling:

- Lateral to metacarpals.

Technique:

- Halfway between carpus and phalanges.

Comments:

- The horizontal beam is accomplished by placing the limb on sponges to have the limb extending straight out from the body. Position the beam, and center on joint as would be done in the ventral position. Typically the metacarpals and digits are radiographed in a single view.

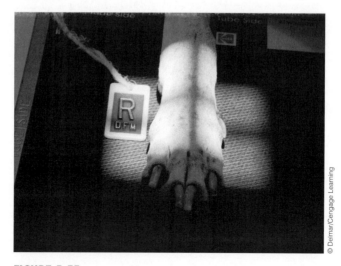

FIGURE 5-55

Proper positioning for the dorsopalmar projection of the metacarpals.

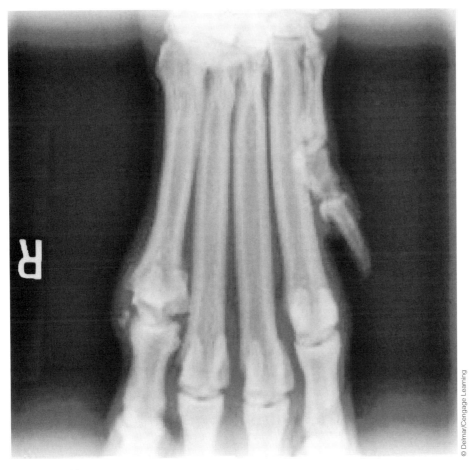

FIGURE 5-56

Dorsopalmar projection of the metacarpals.

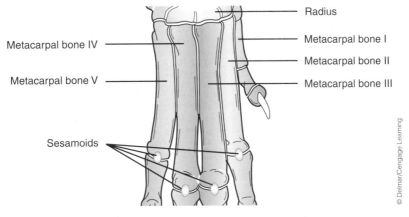

FIGURE 5-57

Anatomical features and landmarks: radius, metacarpal bone I, metacarpal bone II, metacarpal bone III, metacarpal bone IV, metacarpal bone V, and sesamoids.

Lateral Projection of the Metacarpals

Positioning:

- Patient is in lateral recumbency with affected limb down.
- Affected limb is extended down in natural position.
- Sponge is placed under elbow to make the limb even and assist with making metacarpus lateral.

Centering:

- Halfway between carpus and phalanges.

Collimation:

- From the distal third of radius and ulna to the proximal third of the digits.

Labeling:

- Dorsal to metacarpal joint.

Technique:

- Halfway between carpus and phalanges.

Comments:

- Typically the metacarpal and digits are radiographed in a single view.

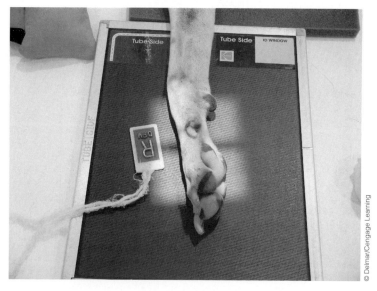

FIGURE 5-58

Proper positioning of the lateral projection of the metacarpals.

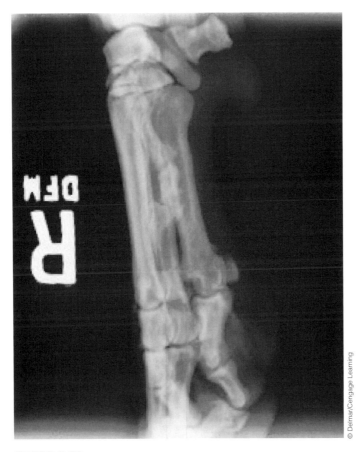

FIGURE 5-59

Lateral projection of the metacarpals.

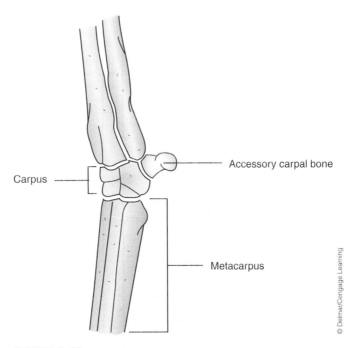

FIGURE 5-60

Anatomical features and landmarks: carpus, accessory carpal bone, and metacarpus.

Dorsopalmar Projection of the Digits

Positioning:

- Patient is in ventral (sternal) recumbency.
- Front legs are extended forward individually.
- Individually tape the medial toe (digit 2) and lateral toe (digit 5), and pull opposite of each other to spread the toes. Alternatively, place cotton balls between each toe.
- Head is extended laterally and to the opposite side of the affected limb, and stabilized with sandbag or tape.
- V-trough can help stabilize caudal half of body.

Centering:

- Center just above digits.

Collimation:

- From the metacarpals proximally to the end of the digits distally.

Labeling:

- Lateral to digits.

Technique:

- Measure midway between metacarpals and digits.

Comments:

- Taping the digits open provides better visualization of the individual bones. If the patient does not have a toe-nail or it is very short, tape around the digit itself. Typically the metacarpal and digits are radiographed in a single view. A horizontal beam projection may also be used.

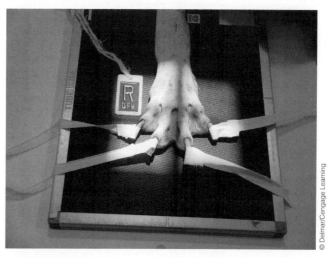

© Delmar/Cengage Learning

FIGURE 5-61

Proper positioning for the dorsopalmar projection of the digits.

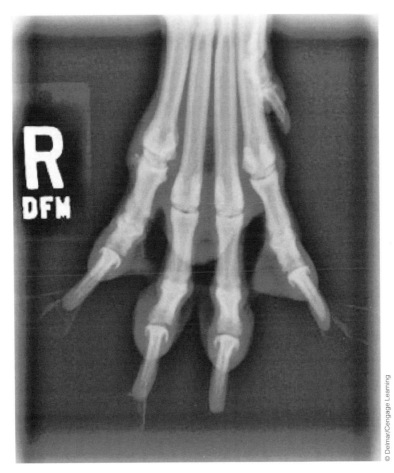

FIGURE 5-62

Dorsopalmar projection of the digits.

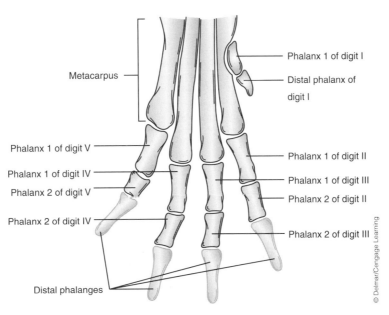

FIGURE 5-63

Anatomical features and landmarks: metacarpus, phalanx 1 of digit I, phalanx 1 of digit II, phalanx 1 of digit III, phalanx 1 of digit IV, phalanx 1 of digit V, phalanx 2 of digit II, phalanx 2 of digit III, phalanx 2 of digit IV, phalanx 2 of digit V, and distal phalanx of digits I, and distal phalanges..

Lateral Projection of the Digits

Positioning:

- Patient is in lateral recumbency with affected limb down.
- Affected limb is extended down in natural position.
- Sponge is placed under elbow to make the limb even and assist with making metacarpus lateral.
- Individually tape the lateral toe (digit 5) and medial toe (digit 2), and pull the lateral toe cranially and the medial toe caudally.

Centering:

- Center just above digits.

Collimation:

- From the metacarpals to the end of the digits.

Labeling:

- Dorsal to the lateral toe (digit 5).

Technique:

- Measure midway between metacarpals and digits.

Comments:

- Taping the digits open provides better visualization of the individual bones. If the patient does not have a toenail or it is very short, tape around the digit itself. Typically the metacarpal and digits are radiographed in a single view.

FIGURE 5-64

Proper positioning for the lateral projection of the digits.

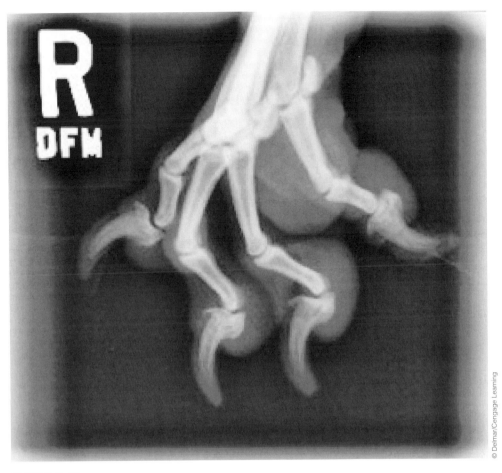

FIGURE 5-65

Lateral projection of the digits.

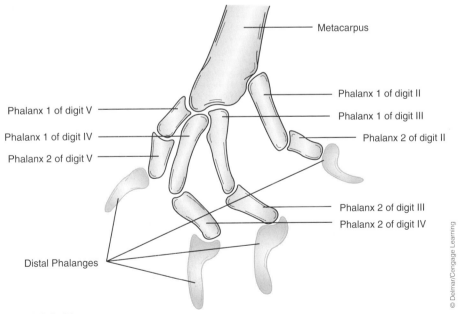

Metacarpus

Phalanx 1 of digit V

Phalanx 1 of digit IV

Phalanx 2 of digit V

Phalanx 1 of digit II

Phalanx 1 of digit III

Phalanx 2 of digit II

Phalanx 2 of digit III

Phalanx 2 of digit IV

Distal Phalanges

FIGURE 5-66

Anatomical features and landmarks: metacarpus, phalanx 1 of digit II, phalanx 1 of digit III, phalanx 1 of digit IV, phalanx 1 of digit V, phalanx 2 of digit II, phalanx 2 of digit III, phalanx 2 of digit IV, phalanx 2 of digit V, and distal phalanges.

CHAPTER 6

HINDLIMB RADIOGRAPHS

OVERVIEW

Radiographic projections of the limbs of the pelvic girdle are often performed to detect fractures. Careful positioning is needed to maintain the limb in a parallel plane against the x-ray cassette to avoid magnification and distortion of the image. The x-ray cassette is normally placed on the tabletop rather than under the table due to the relatively small measurement of dog and cat limbs. Collimation includes joints above and below a bone for images of long bones. Radiographic projections of joints generally include approximately one-third of the bones proximal and distal to the joint. The beam is restricted to just the width needed to include all the necessary structures. This reduces scatter radiation and results in a higher quality image.

When patients are exhibiting signs of pain, a horizontal beam may be used to minimize the need to manipulate the limb. The horizontal beam projection requires supporting the limb on a foam pad and placing the x-ray cassette perpendicular to the tabletop.

Radiographic evaluation of the hindlimbs includes lateral and caudocranial images of the femur, stifle joint, tibia, and fibula. Dorsoplantar and lateral views are usually taken of the tarsus, metatarsus, and phalanges. Oblique views are often needed for the tarsus, and flexed and extended views of the tarsus are routinely performed.

Lateral Projection of the Femur

Positioning:

- Patient is in lateral recumbency with affected limb down.
- Unaffected limb is taped around stifle and tarsus, and abducted out of the way of the affected femoral head.

Centering:

- Midshaft of the femur halfway between the stifle and coxofemoral joint.

Collimation:

- From the ischium on the unaffected limb.

Labeling:

- Label affected limb cranially to stifle.

Technique:

- Measure at the midpart of the femur.

Comments:

- The femur is a more difficult bone in relation to technique because the femoral head is surrounded by thick muscle, and the distal stifle is thin.
- Position the patient with the femoral head toward the cathode end of the x-ray tube.
- With large-muscled dogs, it may be necessary to take two films measuring at both ends.
- Another technique is to place a full-fluid bag over the distal stifle to mimic soft tissue and measure at the femoral head. Note that the edges of the fluid bag will be seen on the film.

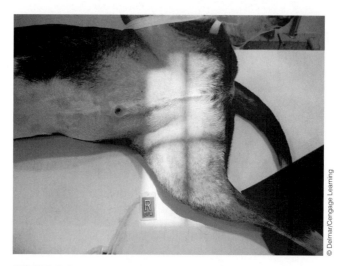

© Delmar/Cengage Learning

FIGURE 6-1

Proper positioning for lateral projection of the femur.

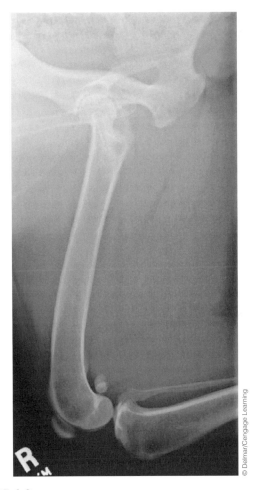

© Delmar/Cengage Learning

FIGURE 6-2

Lateral projection of the femur.

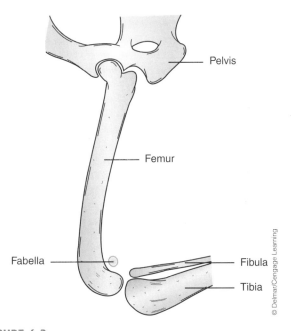

Pelvis

Femur

Fabella

Fibula

Tibia

© Delmar/Cengage Learning

FIGURE 6-3

Anatomical features and landmarks: femur, tibia, fibula, pelvis, and fabella.

Craniocaudal Projection of the Femur

Positioning:

- Patient is in V-trough in dorsal recumbency.
- Tape legs down individually to table.
- Place band of tape around femurs just proximal to the stifles, and pull femurs together to bring the patellas straight over the distal femurs.
- Place sponge under tarsus to avoid rotation of the stifle.

Centering:

- Midshaft of the femur halfway between the stifle and coxofemoral joint.

Collimation:

- Cranial: from the greater trochanter to the proximal third of the tibia.
- Lateral: from the ventral midline to the body wall.

Labeling:

- Place label lateral to the body proximally or distally.

Technique:

- Measure at the midportion of the femur.

Comments:

- Measure the femur itself without compressing the muscles. Avoid including the space between the femur and tabletop because this will result in overexposure of the film.

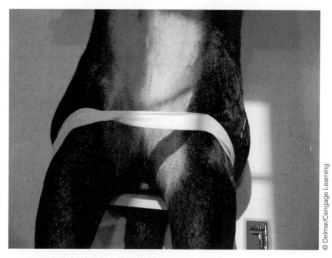

FIGURE 6-4

Proper positioning for craniocaudal projection of the femur.

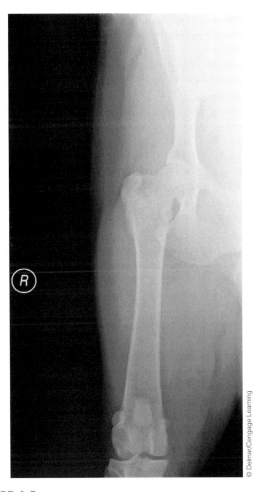

FIGURE 6-5

Craniocaudal projection of the femur.

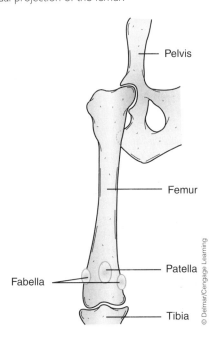

FIGURE 6-6

Anatomical features and landmarks: femur, patella, tibia, pelvis, and fabella.

Caudocranial Projection of the Stifle

Positioning:

- Patient is in V-trough in ventral recumbency.
- Pull both hindlimbs behind patient and place band of tape around femurs just proximal to the stifles, and pull femurs together to bring the patellas straight over the distal femurs.
- Alternatively, flex the unaffected limb up next to the body if patient is more comfortable.

Centering:

- Center on stifle joint.

Collimation:

- Proximal to the distal third of the femur and distal to proximal third of the tibia.

Labeling:

- Place label lateral to the joint.

Technique:

- Measure at the center of the joint.

Comments:

- It is preferable to angle the x-ray tube head cranially approximately 10–15 degrees to obtain an image through the stifle joint.
- The horizontal beam technique may also be useful (Figure 6-8). Place the limb on sponges so that the limb extends straight from the body. Position the beam, and center on joint as you would in the ventral position.

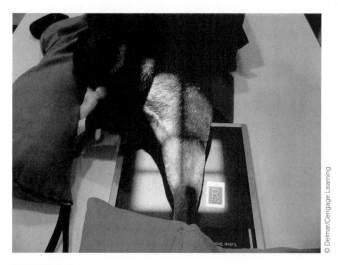

FIGURE 6-7

Proper positioning for caudocranial projection of the stifle.

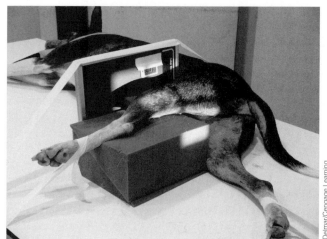

FIGURE 6-8

The horizontal beam technique for the caudocranial projection of the stifle.

© Delmar/Cengage Learning

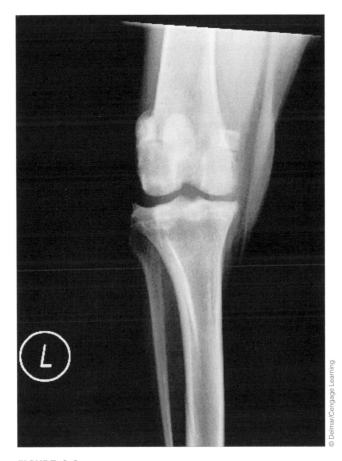

FIGURE 6-9

Caudocranial projection of the stifle.

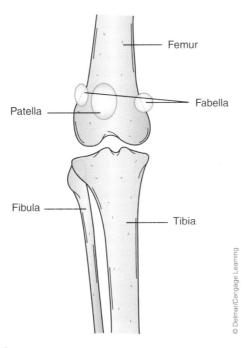

FIGURE 6-10

Anatomical features and landmarks: femur, patella, tibia, fibula, and fabella.

Lateral Projection of the Stifle

Positioning:

- Patient is in lateral recumbency with affected limb down.
- Unaffected limb is taped around stifle and tarsus, and pulled up and out of the way.

Centering:

- Center on stifle joint.

Collimation:

- Proximal to the distal third of the femur and distal to proximal third of the tibia.

Labeling:

- Place label cranial to joint.

Technique:

- Measure the center of the joint. Palpate indentation of joint (fat pad) on the dorsal aspect of stifle. Center on indentation.

Comments:

- It is important to include the soft tissues on the caudal aspect of the stifle joint to visualize the fascial stripe that runs proximally and distally. When this stripe is absent, it is a sign of joint effusion.

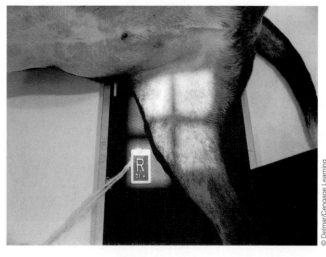

FIGURE 6-11

Proper positioning for lateral projection of the stifle.

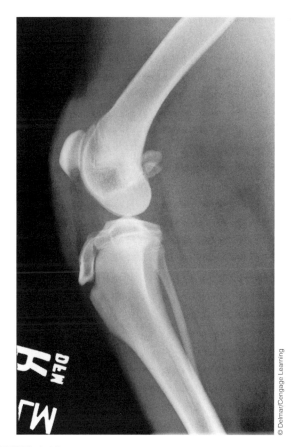

FIGURE 6-12

Lateral projection of the stifle.

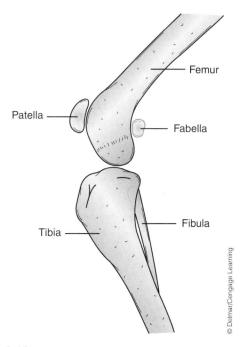

FIGURE 6-13

Anatomical features and landmarks: femur, patella, fabella, tibia, and fibula.

Caudocranial Projection of the Tibia and Fibula

Positioning:

- Patient is in V-trough in ventral recumbency.
- Pull both hind limbs behind patient and place band of tape around femurs just proximal to the stifles, and pull femurs together to bring the patellas straight over the distal femurs.
- Alternatively, flex the unaffected limb up next to the body if patient is more comfortable.

Centering:

- Center on the midshaft of the tibia halfway between the stifle and tarsus.

Collimation:

- From proximal to the distal third of the femur to distal to the proximal third of the tarsus.

Labeling:

- Place label on the lateral side of the tibia.

Technique:

- Measure midshaft of the tibia.

Comments:

- The horizontal beam technique may also be useful (Figure 6-15). Place the limb on sponges so that the limb extends straight out from the body. Position the beam, and center on joint as would be done in the ventral position.

© Delmar/Cengage Learning

FIGURE 6-14

Proper positioning for caudocranial projection of the tibia and fibula.

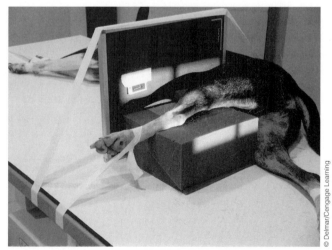

© Delmar/Cengage Learning

FIGURE 6-15

The horizontal beam technique for the caudocranial projection of the tibia and fibula.

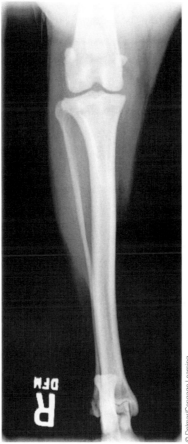

FIGURE 6-16

Caudocranial projection of the tibia and fibula.

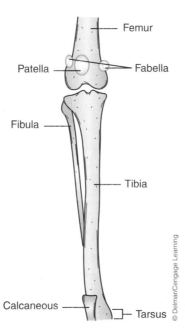

FIGURE 6-17

Anatomical features and landmarks: femur, patella, fabella, tibia, fibula, calcaneous, and tarsus.

Lateral Projection of the Tibia

Positioning:

- Patient is in lateral recumbency with affected limb down.
- Unaffected limb is taped around stifle and tarsus, and pulled up and out of the way.

Centering:

- Midshaft of the tibia halfway between the stifle and tarsus.

Collimation:

- From proximal to the distal third of the femur to distal to the proximal third of the tarsus.

Labeling:

- Place label cranial.

Technique:

- Measure midshaft of tibia.

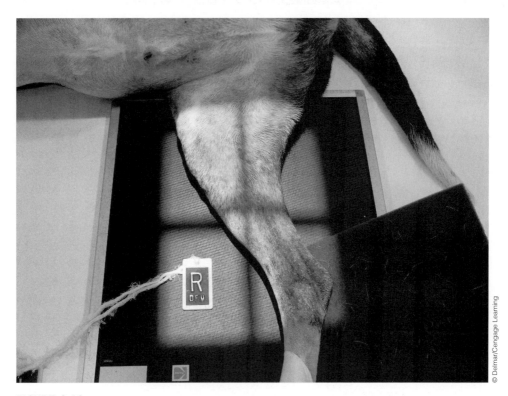

FIGURE 6-18

Proper positioning for lateral projection of the tibia.

© Delmar/Cengage Learning

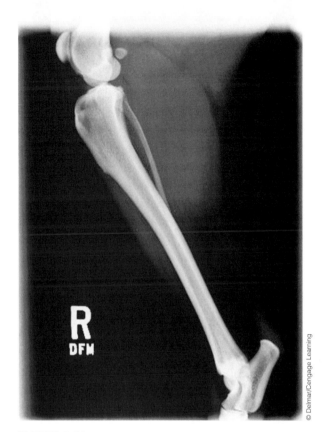

FIGURE 6-19

Lateral projection of the tibia.

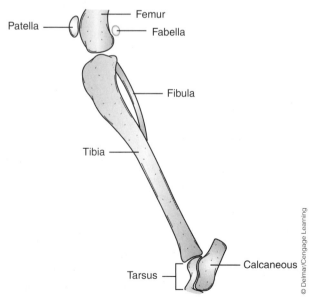

FIGURE 6-20

Anatomical features and landmarks: femur, patella, fabella, tibia, fibula, calcaneous, and tarsus.

Dorsoplantar Projection of the Tarsus

Positioning:

- Patient is in dorsal recumbency in the V-trough.
- Place sponges under tarsus to bring cassette close to the tarsus.
- Tape affected tarsus to table.

Centering:

- Center on tarsal joint halfway between the distal tibia and the proximal metatarsals.

Collimation:

- Proximal to the distal third of the tibia and distal to the proximal third of the metatarsals.

Labeling:

- Place label lateral to the joint.

Technique:

- Measure the center of the joint.

FIGURE 6-21

Proper positioning for dorsoplantar projection of the tarsus.

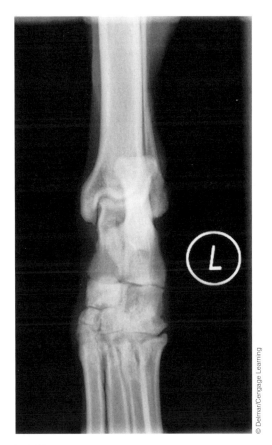

FIGURE 6-22

Dorsoplantar projection of the tarsus.

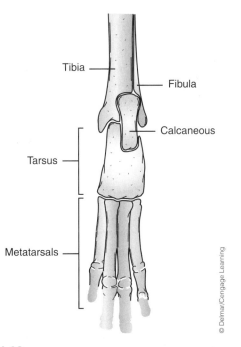

FIGURE 6-23

Anatomical features and landmarks: tibia, fibula, calcaneous, tarsus, and metatarsals.

Lateral Projection of the Tarsus

Positioning:

- Patient is in lateral recumbency with affected limb down.
- Unaffected limb is taped around stifle and tarsus and dorsally.

Centering:

- Center on tarsal joint halfway between the distal tibia and the proximal metatarsals.

Collimation:

- Proximal to the distal third of the tibia and distal to the proximal third of the metatarsals.

Labeling:

- Place label cranial to the joint.

Technique:

- Measure the center of the joint.

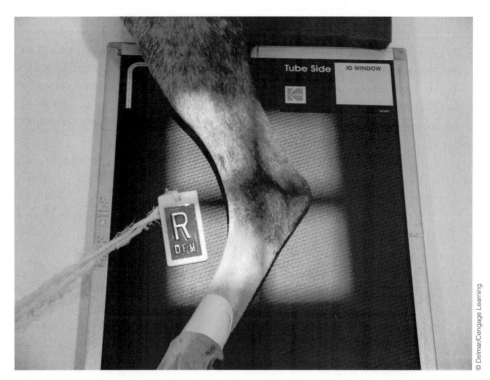

FIGURE 6-24

Proper positioning for lateral projection of the tarsus.

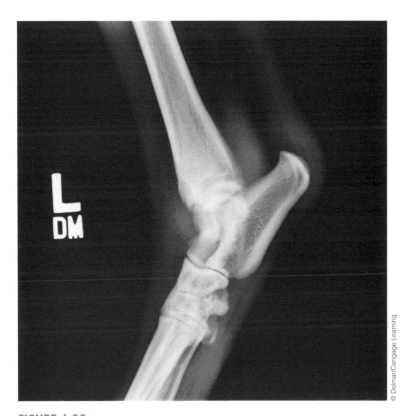

© Delmar/Cengage Learning

FIGURE 6-25

Lateral projection of the tarsus.

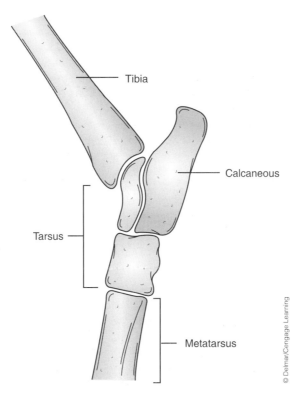

© Delmar/Cengage Learning

FIGURE 6-26

Anatomical features and landmarks: tibia, calcaneous, tarsus, and metatarsus.

Flexed Lateral Projection of the Tarsus

Positioning:

- Patient is in lateral recumbency with affected limb down.
- Unaffected limb is taped around stifle and tarsus dorsally.
- Tape around caudal tibia and proximal metatarsus to achieve full flexion of the tarsus.

Centering:

- Center on tarsal joint between the stifle and metatarsals.

Collimation:

- From the distal portion of the tibia to the proximal portion of the metatarsals.

Labeling:

- Place label cranial to the joint.

Technique:

- Measure to the center of the joint.

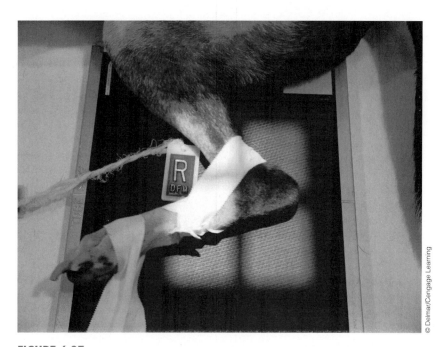

© Delmar/Cengage Learning

FIGURE 6-27

Proper positioning for flexed lateral projection of the tarsus.

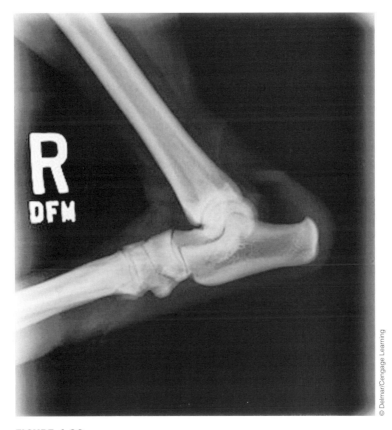

FIGURE 6-28

Flexed lateral projection of the tarsus.

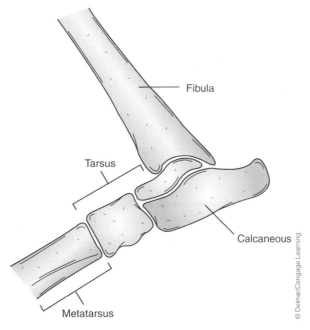

FIGURE 6-29

Anatomical features and landmarks: fibula, calcaneous, tarsus, and metatarsus.

Extended Lateral Projection of the Tarsus

Positioning:

- Patient is in lateral recumbency with affected limb down.
- Unaffected limb is taped around stifle and tarsus, and pulled dorsally.
- Pull affected limb straight out away from body for full extension.

Centering:

- Center on tarsal joint between the stifle and metatarsals.

Collimation:

- From the distal portion of the tibia to the proximal portion of the metatarsals.

Labeling:

- Place label cranial to the joint.

Technique:

- Measure the center of the joint.

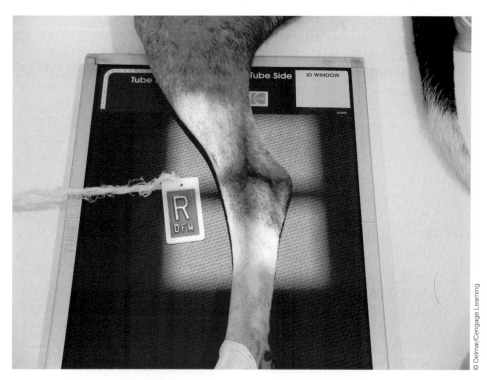

© Delmar/Cengage Learning

FIGURE 6-30

Proper positioning for extended lateral projection of the tarsus.

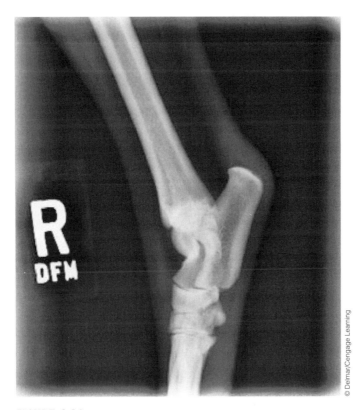

FIGURE 6-31

Extended lateral projection of the tarsus.

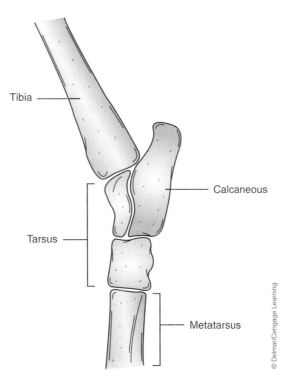

FIGURE 6-32

Anatomical features and landmarks: tibia, calcaneous, tarsus, and metatarsus.

Dorsolateral/Plantaromedial Oblique Projection of the Tarsus

Positioning:

- Patient is in dorsal recumbency in the V-trough.
- Place sponges under tarsus to bring cassette close to the tarsus.
- Tape affected tarsus to table.
- Tilt patient from you approximately 15–20 degrees to achieve an oblique angle of the tarsus with the medial side facing the tube.

Centering:

- Center on tarsal joint.

Collimation:

- From the distal portion of the tibia to the proximal portion of the metatarsals.

Labeling:

- Place label lateral to the joint.

Technique:

- Measure at center of tarsal joint.

Comments:

- If x-ray machine is capable, use standard dorsoplantar position, and angle the x-ray head 15–20 degrees toward the medial side of the tarsus.

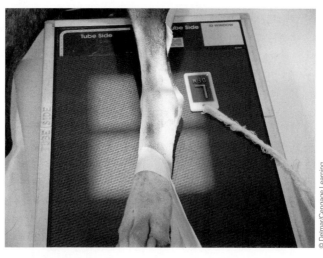

FIGURE 6-33

Proper positioning for dorsolateral/plantaromedial oblique projection of the tarsus.

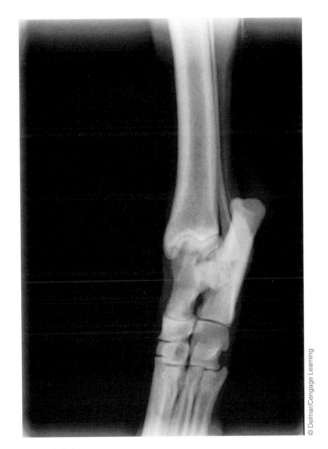

FIGURE 6-34

Dorsolateral/plantaromedial oblique projection of the tarsus.

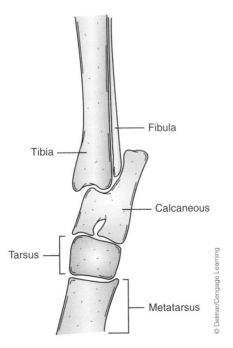

FIGURE 6-35

Anatomical features and landmarks: tibia, fibula, calcaneous, tarsus, and metatarsus.

Dorsomedial/Plantarolateral Oblique Projection of the Tarsus

Positioning:

- Patient is in dorsal recumbency in the V-trough.
- Place sponges under tarsus to bring cassette close to the tarsus.
- Tape affected tarsus to table.
- Tilt patient approximately 15–20 degrees to achieve an oblique angle of the tarsus with the lateral side facing the tube.

Centering:

- Center on tarsal joint.

Collimation:

- From the distal portion of the tibia to the proximal portion of the metatarsals.

Labeling:

- Place label lateral to the joint.

Technique:

- Measure at center of tarsal joint.

Comments:

- If x-ray machine is capable, use standard dorsoplantar position, and angle the x-ray head 15–20 degrees toward the lateral side of the tarsus.

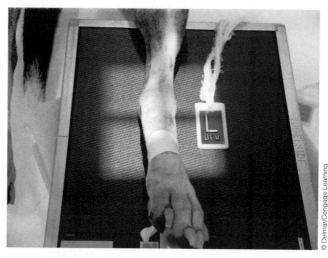

FIGURE 6-36

Proper positioning for dorsomedial/plantarolateral oblique projection of the tarsus.

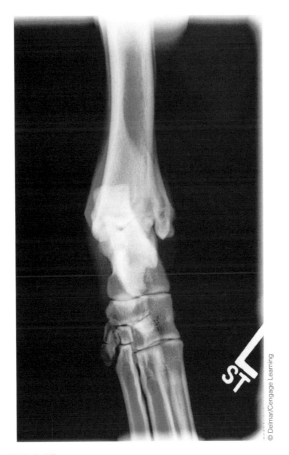

FIGURE 6-37

Dorsomedial/plantarolateral oblique projection of the tarsus.

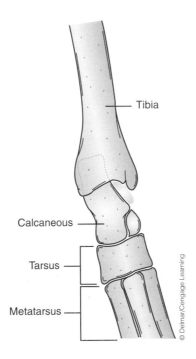

FIGURE 6-38

Anatomical features and landmarks: tibia, calcaneous, tarsus, and metatarsus.

Dorsoplantar Projection of the Metatarsus

Positioning:

- Patient is in dorsal recumbency in the V-trough.
- Place sponges under metatarsus to bring cassette close to the metatarsus.
- Tape affected metatarsus to table.

Centering:

- Center on the metatarsal joint.

Collimation:

- From the distal portion of the tarsus to the proximal portion of the phalanges. Collimation through the distal phalanges is also acceptable.

Labeling:

- Place label lateral to the joint.

Technique:

- Measure the center of the joint.

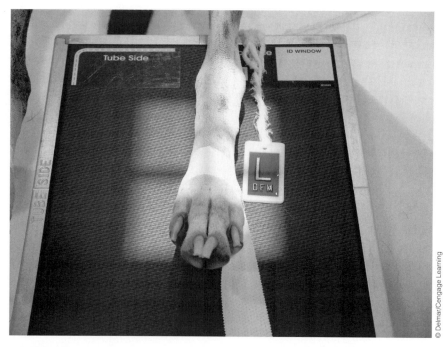

FIGURE 6-39

Proper positioning for dorsoplantar projection of the metatarsus.

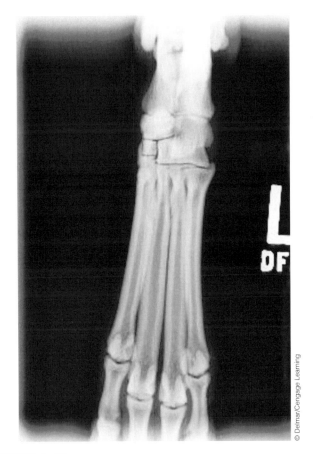

FIGURE 6-40

Dorsoplantar projection of the metatarsus.

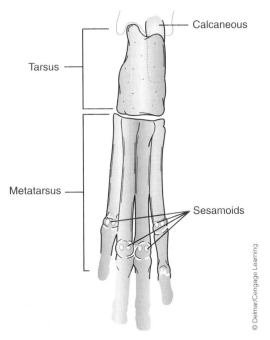

FIGURE 6-41

Anatomical features and landmarks: calcaneous, tarsus, metatarsus, and sesamoids.

Lateral Projection of the Metatarsus

Positioning:

- Patient is in lateral recumbency with affected limb down.
- Unaffected limb is taped around stifle and tarsus, and pulled up and out of the way.

Centering:

- Center on metatarsal joint.

Collimation:

- From the distal portion of the tarsus to the proximal portion of the phalanges.

Labeling:

- Place label cranial to the joint.

Technique:

- Measure the center of the joint.

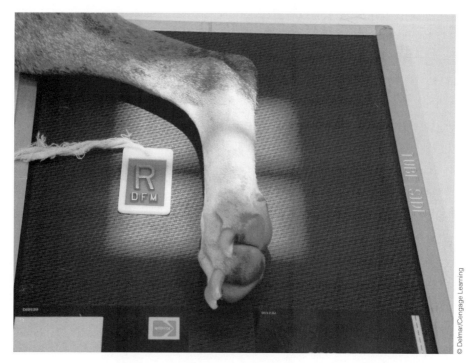

FIGURE 6-42

Proper positioning for lateral projection of the metatarsus.

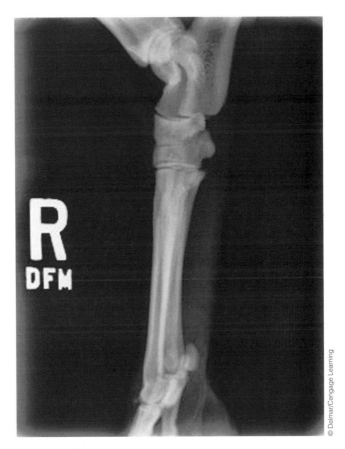

FIGURE 6-43

Lateral projection of the metatarsus.

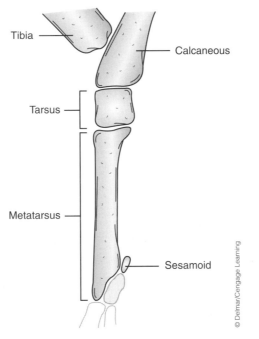

FIGURE 6-44

Anatomical features and landmarks: tibia, calcaneous, tarsus, metatarsus, and sesamoid.

Dorsopalmar Projection of the Digits

Positioning:

- Patient is in dorsal recumbency.
- Hindlimbs are pulled caudally individually.
- Individually tape the medial toe (digit 2) and lateral toe (digit 5), and pull opposite of each other to spread the toes. Alternatively place cotton balls between each toe.
- V-trough can help stabilize cranial half of body.

Centering:

- Center just above digits.

Collimation:

- Include metatarsus through the distal end of the digits.

Labeling:

- Lateral to digits.

Technique:

- Measure midway between metatarsus and digits.

Comments:

- Taping the digits open provides better visualization of the individual bones. If the patient does not have a toenail or it is very short, tape around the digit itself.
- Typically, the metatarsus and digits are radiographed in a single view.
- The horizontal beam technique may also be useful.

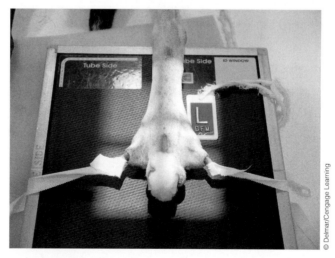

FIGURE 6-45

Proper positioning for dorsopalmar projection of the digits.

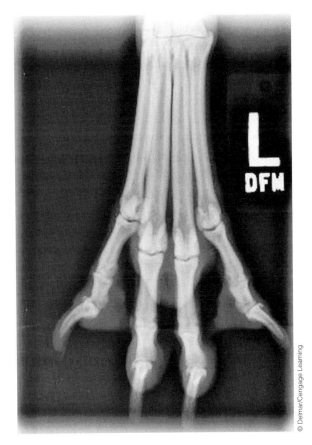

FIGURE 6-46

Dorsopalmar projection of the digits.

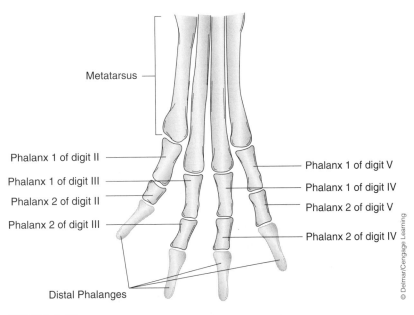

FIGURE 6-47

Anatomical features and landmarks: metatarsus, phalanx 1 of digit V, phalanx 1 of digit IV, phalanx 1 of digit III, phalanx 1 of digit II, phalanx 2 of digit V, phalanx 2 of digit IV, phalanx 2 of digit III, phalanx 2 of digit II, and distal phalanges.

Lateral Projection of the Digits

Positioning:

- Patient is in lateral recumbency with affected limb down.
- Affected limb is pulled down in natural position.
- Sponge is placed under stifle to make the limb even and assist with making metatarsus lateral.
- Individually tape the lateral toe (digit 5) and medial toe (digit 2), and pull the lateral toe cranially and the medial toe caudally.

Centering:

- Center just above digits.

Collimation:

- Include all the metatarsal and digits.

Labeling:

- Cranial to label the lateral toe (digit 5).

Technique:

- Measure midway between metatarsus and digits.

Comments:

- Taping the digits open provides better visualization of the individual bones. If the patient does not have a toe-nail or it is very short, tape around the digit itself.
- Typically the metatarsus and digits are radiographed in a single view.

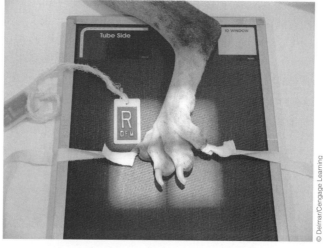

FIGURE 6-48

Proper positioning for lateral projection of the digits.

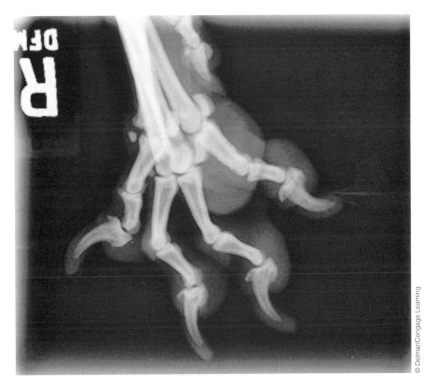

FIGURE 6-49

Lateral projection of the digits.

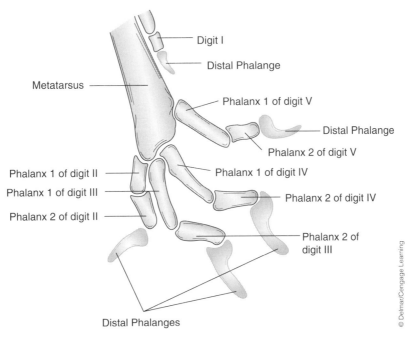

FIGURE 6-50

Anatomical features and landmarks: metatarsus, phalanx 1 of digit II, phalanx 1 of digit III, phalanx 1 of digit IV, phalanx 1 of digit V, digit I, phalanx 2 of digit II, phalanx 2 of digit III, phalanx 2 of digit IV, phalanx 2 of digit V, and distal phalanges.

CHAPTER 7

SKULL RADIOGRAPHS

OVERVIEW

Indications for examination of skull radiographs include evaluation of the tympanic bullae, nasal sinuses, and foramen magnum, as well as evaluation to detect bone lesions. The patient must be positioned precisely because evaluation often focuses on symmetry between the two sides of the skull. Nearly all patients require general anesthesia for proper positioning. The endotracheal tube may need to be removed or repositioned before obtaining the exposure depending on the purpose of the radiograph.

For most small animals, the skull is approximately the same width both laterally and dorsoventrally. Most skull radiographs use a single measurement representing the widest area of the cranium. Measurements for radiographs of the nasal passages are taken at a point slightly rostral to the widest area of the cranium to avoid overexposing the air-filled sinuses.

For most routine radiographic studies of the skull, a lateral projection and either a ventrodorsal (VD) or dorsoventral (DV) view are obtained. For evaluation of the tympanic bullae, a DV view, right and left lateral oblique views, and an open mouth view are obtained. Evaluation of the nasal passages commonly requires a lateral view, either a DV or VD view, a frontal (rostrocaudal) view, and an open mouth view. Commonly used projections for evaluation of the temporomandibular joint include right and left lateral oblique views and a DV view. Although the entire skull may be included for most views, to enhance detail, the radiograph should be tightly collimated to the primary area of interest to reduce scatter radiation and provide a higher quality image.

The following pages illustrate the proper positioning and technique for skull radiographs.

Lateral Projection of the Skull

Positioning:

- Right or left lateral recumbency with the affected side toward the cassette.
- Foam pads placed under the mandible to maintain the sagittal plane of the skull in a position parallel to the x-ray cassette.

Centering:

- Lateral canthus of the eye socket.

Collimation:

- Occipital protuberance to the tip of the nose.
- Dorsal right and ventral skull fully within collimated area.

Labeling:

- Mark the side against the x-ray cassette as either left or right.

Technique:

- Measure at the widest point of the cranium.

Comments:

- The rami of the mandible and tympanic bullae are superimposed on the finished radiograph. If nontraumatized and present, the eyes and canine teeth can be a useful landmark in assisting with the lateral projection. Position the skull so that the eyes or canine teeth are even and parallel with the cassette.

FIGURE 7-1

Proper positioning for lateral projection of the skull.

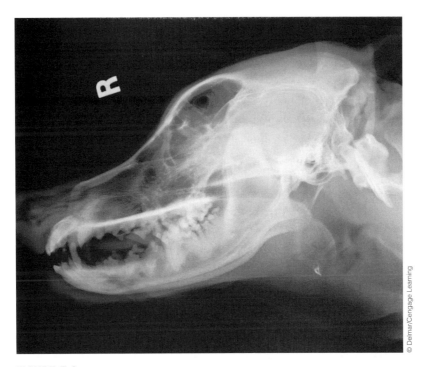

FIGURE 7-2

Lateral projection of the skull.

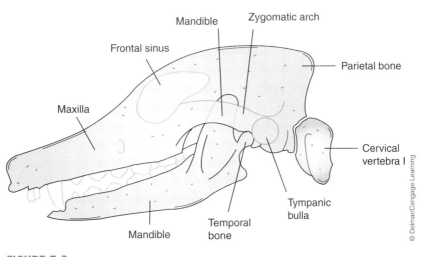

FIGURE 7-3

Anatomical features and landmarks: tympanic bulla, temporal bone, mandible, zygomatic arch, parietal bone, frontal sinus, maxilla, and cervical vertebra I.

DV Projection of the Skull

Positioning:

- Sternal recumbency.
- Sandbag placed across the cervical region to maintain placement of the head against the x-ray cassette.
- Tape can be used across the maxilla to maintain vertical alignment of the head on the x-ray cassette.

Centering:

- Midway between the tip of the nose to just caudal to the occipital protuberance at the base.

Collimation:

- Occipital protuberance to the tip of the nose.
- Zygomatic arches fully within collimated area.

Labeling:

- Mark either left or right dependent side, or both.

Technique:

- Measure at the widest point of the cranium just caudal to the orbit.

FIGURE 7-4

Proper positioning for DV projection of the skull.

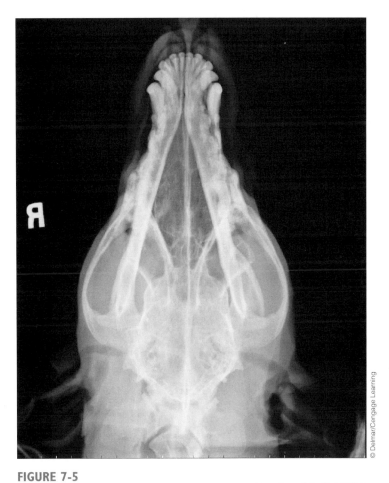

FIGURE 7-5

DV projection of the skull.

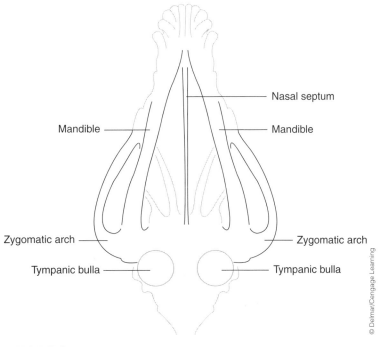

FIGURE 7-6

Anatomical features and landmarks: tympanic bulla, zygomatic arch, mandible, and nasal septum.

VD Projection of the Skull

Positioning:

- Dorsal recumbency.
- Foam pad or sandbag is positioned under neck to maintain hard palate in parallel alignment with x-ray cassette.
- A V-trough can be used to aid in maintaining vertical alignment.
- Forelimbs are secured caudally.
- Foam pad under head or tape across mandibles to avoid rotation of head.

Centering:

- Midway between the tip of the nose to just caudal to the occipital protuberance at the base.

Collimation:

- Occipital protuberance to the tip of the nose.
- Zygomatic arches fully within collimated area.

Labeling:

- Mark either left or right dependent side, or both.

Technique:

- Measure at the widest point of the cranium just caudal to the orbit.

Comments:

- This view should be included in a nasal series because the nasal passages are on the dorsal portion of the skull and closest to the cassette.

FIGURE 7-7

Proper positioning for VD projection of the skull.

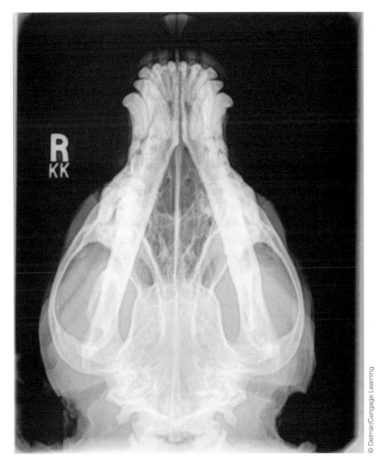

FIGURE 7-8

VD projection of the skull.

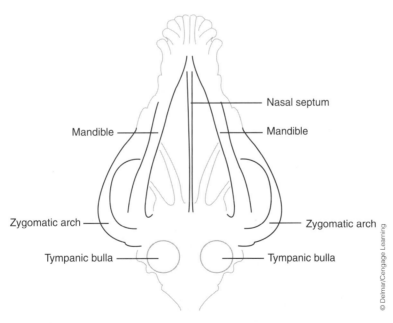

FIGURE 7-9

Anatomical features and landmarks: tympanic bulla, zygomatic arch, mandible, and nasal septum.

Rostrocaudal Sinuses Closed Mouth Projection

Positioning:

- Dorsal recumbency.
- Foam pad or sandbag is positioned under neck.
- A V-trough can be used to aid in maintaining vertical alignment.
- Forelimbs are secured caudally.
- Tape or gauze to direct nose caudally to maintain hard palate perpendicular to the x-ray cassette and parallel to the x-ray beam.

Centering:

- Between the eyes.

Collimation:

- Include all of occipital crest to dorsal aspect of the nasal planum.
- Zygomatic arches fully within collimated area.

Labeling:

- Mark either left or right dependent side, or both.

Technique:

- Measure midpoint at the level of the eyes.

Comments:

- This view is commonly referred to as the "skyline" view.

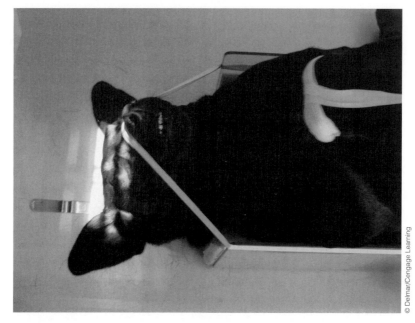

© Delmar/Cengage Learning

FIGURE 7-10

Proper positioning for rostrocaudal sinuses closed mouth projection.

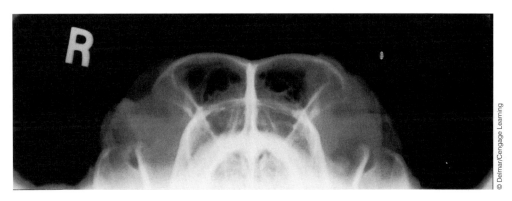

FIGURE 7-11

Rostrocaudal sinuses closed mouth projection.

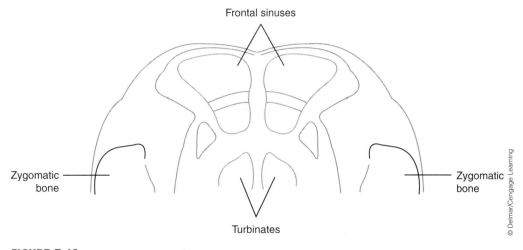

FIGURE 7-12

Anatomical features and landmarks: turbinates, frontal sinuses, and zygomatic bone.

Rostrocaudal Foramen Magnum Projection

Positioning:

- Dorsal recumbency.
- Foam pad or sandbag is positioned under neck.
- A V-trough can be used to aid in maintaining vertical alignment.
- Forelimbs are secured caudally.
- Tape or gauze to direct nose caudally approximately 30 degrees with the mandible close to the chest.

Centering:

- Between the eyes.

Collimation:

- Include all of occipital crest to tympanic bullae.
- Zygomatic arches fully within collimated area.

Labeling:

- Mark either left or right dependent side, or both.

Technique:

- Measure midpoint at the level of the eyes.

Comments:

- This view is also referred to as the "keyhole" or "town-crown" view.

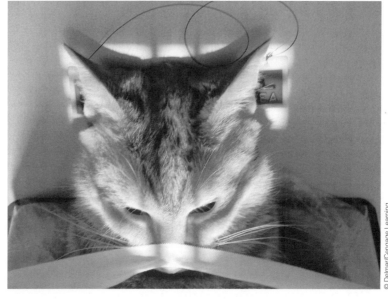

FIGURE 7-13

Proper positioning for rostrocaudal foramen magnum projection.

© Delmar/Cengage Learning

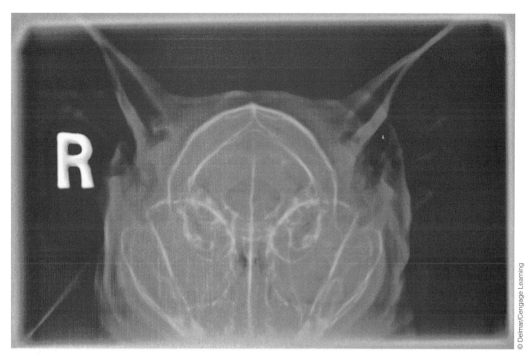

FIGURE 7-14

Rostrocaudal foramen magnum projection.

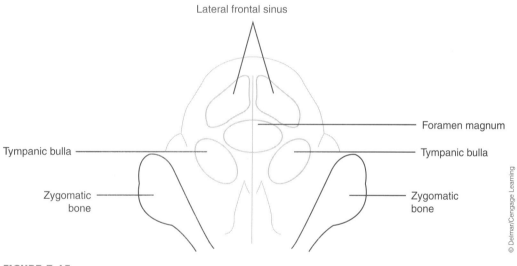

FIGURE 7-15

Anatomical features and landmarks: tympanic bulla, foramen magnum, zygomatic bone, and lateral frontal sinus.

VD Rostrocaudal Nasal Open Mouth Projection

Positioning:

- Dorsal recumbency.
- With the skull placed outside, place caudal portion of patient in V-trough to aid in keeping the skull and body straight.
- Position head so that the hard palate is in parallel alignment with x-ray cassette, tape across maxillary canines, and secure to the sides of the table.
- Place tape across or behind mandibular canines, and pull mandible along with the tongue and endotracheal tube to open the mouth. Secure tape to the sides of the V-trough.
- Pull forelimbs and secure caudally.

Centering:

- Tilt tube head caudally approximately 15 degrees, centering on the back of the palate.
- If using the cassette/Bucky tray, line up tray and collimator light to account for the tilt.

Collimation:

- Horizontally inside the zygomatic arches, and vertically from the tip of the maxilla to the back of the palate, beyond where the mandible will superimpose over it.

Labeling:

- Mark either left or right dependent side, or both.

Technique:

- Measure at the thickest area near the commissure of the lip.
- Angling the beam may not be necessary in feline patients, or the x-ray beam may be directed at a 5–10 degree angle.

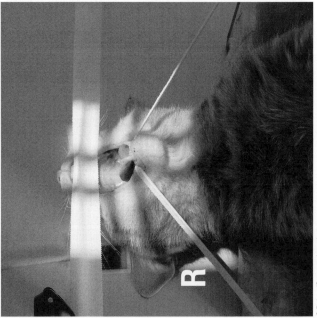

FIGURE 7-16

Proper positioning for VD rostrocaudal nasal open mouth projection.

© Delmar/Cengage Learning

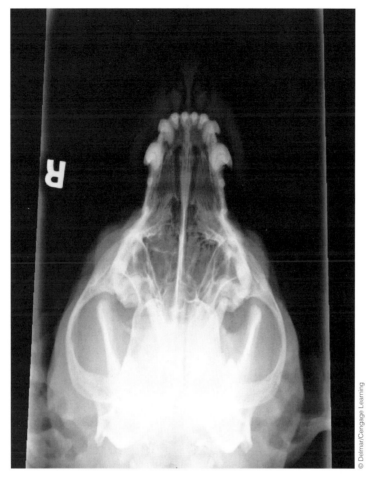

FIGURE 7-17

VD rostrocaudal nasal open mouth projection

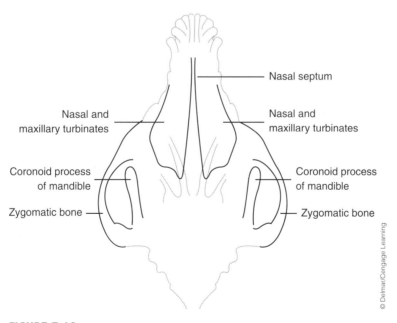

FIGURE 7-18

Anatomical features and landmarks: nasal and maxillary turbinates, zygomatic bone, nasal septum, and coronoid process of mandible.

Rostrocaudal Tympanic Bullae Open Mouth Projection

Positioning:

- Dorsal recumbency.
- Foam pad or sandbag is positioned under neck.
- A V-trough can be used to aid in maintaining vertical alignment.
- Forelimbs are secured caudally.
- Tape or gauze around maxilla to pull nose approximately 10 degrees cranially.
- Tape or gauze around mandible to pull mandible approximately 10 degrees caudally.

Centering:

- Beam is centered just above the base of the tongue and just below the soft palate approximately at the commissure of the mouth.

Collimation:

- Zygomatic arches fully within collimated area with enough lateral to include the marker.

Labeling:

- Mark either left or right dependent side, or both.

Technique:

- Measure at the commissure of the mouth.

Comments:

- A plastic-mouth speculum may also be used to maintain the open mouth position, taking care not to allow the weight of the speculum to rotate the head.
- A 1 mL syringe barrel with the plunger removed works well if you do not have a speculum. Cut both ends of the barrel (nail trimmers work well for this), and place open ends on the upper and lower canine teeth. Length can be modified accordingly.
- The rostrocaudal projection in cats can also be taken with the mouth closed and the skull tipped cranially approximately 10 degrees (Figures 7-20, 7-23, and 7-24). This is due to the fact that cat bullae are anatomically farther caudal on the skull than dog bullae.

FIGURE 7-19

Proper positioning for rostrocaudal tympanic bullae open mouth projection.

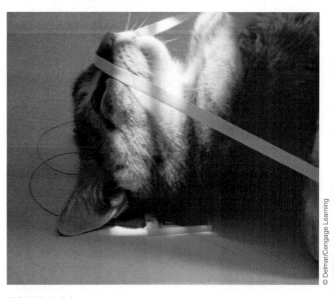

FIGURE 7-20

Proper positioning for rostrocaudal projection in cats.

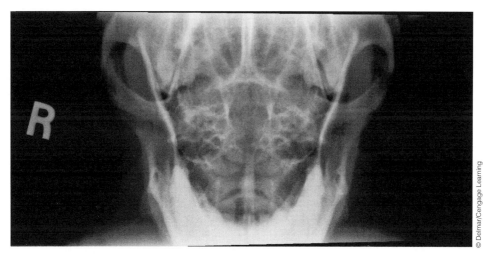

© Delmar/Cengage Learning

FIGURE 7-21

Rostrocaudal tympanic bullae open mouth projection.

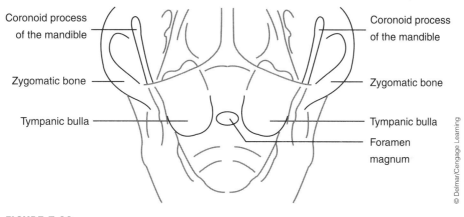

© Delmar/Cengage Learning

FIGURE 7-22

Anatomical features and landmarks: tympanic bulla, foramen magnum, coronoid process of the mandible, and zygomatic bone.

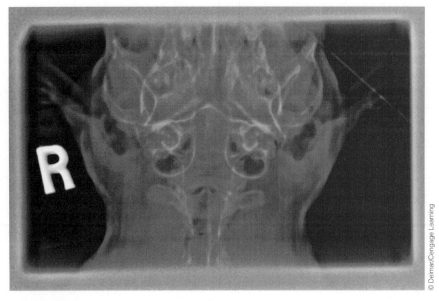

FIGURE 7-23

Rostrocaudal projection in cats.

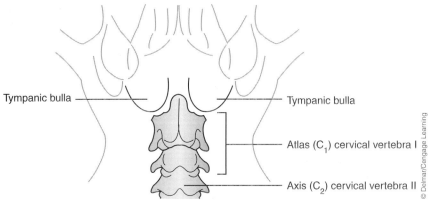

FIGURE 7-24

Anatomical features and landmarks: tympanic bulla, atlas (C_1) cervical vertebra I, and axis (C_2) cervical vertebra II.

DV Tympanic Bullae Projection

Positioning:

- Sternal recumbency.
- Sandbag is placed across the cervical region to maintain placement of the head against the x-ray cassette.
- Tape can be used across the maxilla to maintain vertical alignment of the head on the x-ray cassette.

Centering:

- Palpate the base of the ear to determine placement on the skull. Center the beam on the dorsal midline of the skull between the ears.

Collimation:

- Palpate the ear, and collimate just cranial and caudal to the ear.
- Right and left side of skull is fully within collimated area.

Labeling:

- Mark either left or right dependent side, or both.

Technique:

- Measure at the widest point of the cranium just caudal to the orbit and over the bullae.

Comments:

- DV position is preferred due to the bullae being closer to the cassette.

FIGURE 7-25

Proper positioning for DV tympanic bullae projection.

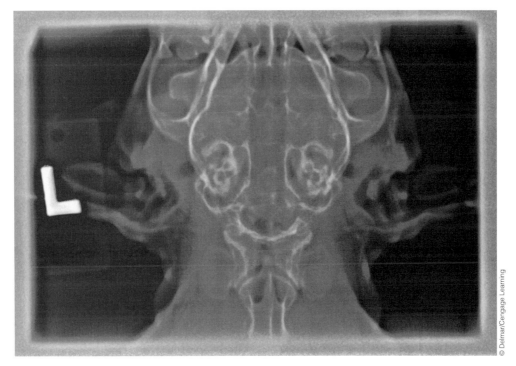

FIGURE 7-26

DV tympanic bullae projection.

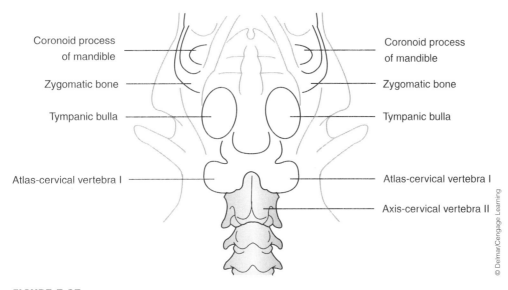

Coronoid process of mandible

Zygomatic bone

Tympanic bulla

Atlas-cervical vertebra I

Coronoid process of mandible

Zygomatic bone

Tympanic bulla

Atlas-cervical vertebra I

Axis-cervical vertebra II

FIGURE 7-27

Anatomical features and landmarks: tympanic bulla, zygomatic bone, coronoid process of mandible, atlas-cervical vertebra I, and axis-cervical vertebra II.

Lateral Oblique Tympanic Bullae Projection

Positioning:

- Place patient in lateral recumbency.
  - Right lateral recumbency for left oblique bulla.
  - Left lateral recumbency for the right bulla.
- From the lateral position, allow the skull to oblique itself naturally down toward the table approximately 30–40 degrees. This projects the down bulla and keeps it from being superimposed over the skull.

Centering:

- Palpate and center over the ear to include both dorsal and ventral skull.

Collimation:

- Slightly cranial and caudal to the ear.

Labeling:

- Label the projected bulla.
- The patient in right lateral recumbency will have the right marker dorsal and the left marker ventral. The patient in left lateral recumbency will have the left marker dorsal and the right marker ventral.

Technique:

- Measure at the widest point of the cranium just caudal to the orbit and over the bullae.

Comments:

- The projection should have the dependent bulla isolated and not superimposed over the skull.
- This oblique projection can also be used to obtain images of the mandible or maxilla by collimating to include the entire mandible or maxilla. Both right and left lateral oblique views are required.

© Delmar/Cengage Learning

FIGURE 7-28

Proper positioning for lateral oblique tympanic bullae projection.

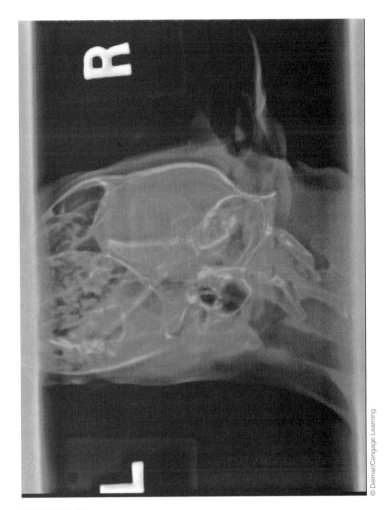

FIGURE 7-29

Lateral oblique tympanic bullae projection.

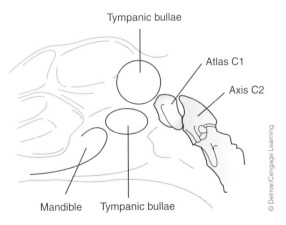

FIGURE 7-30

Anatomical features and landmarks: tympanic bulla, atlas-cervical vertebra, and mandible.

Lateral Oblique Temporomandibular Joint Projection

Positioning:

- Place patient in lateral recumbency; normally, both right and left lateral oblique views are obtained from each patient.
  - Right lateral recumbency for the right temporomandibular joint.
  - Left lateral recumbency for the left temporomandibular joint.
- From the lateral position, allow the skull to oblique itself naturally down toward the table approximately 10 degrees. This projects the temporomandibular joint down and keeps it from being superimposed over the skull.
- Place a small sponge to lift the rostral part of the skull (at the nose) up approximately 10 degrees project the temporomandibular joint rostral and keep it from being superimposed over the skull.

Centering:

- The beam is centered just cranial to the ear or bulla.

Collimation:

- Collimate just cranial and caudal to the joint.

Labeling:

- Label the dependent temporomandibular joint.

Technique:

- Measure just caudal to the orbit in front of the ear over the joint.

Comments:

- When evaluating for potential luxation, an open mouth oblique view of each temporomandibular joint might be required.

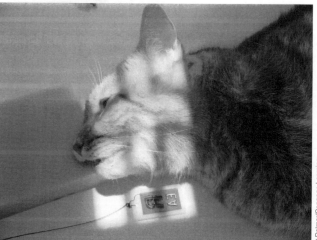

© Delmar/Cengage Learning

FIGURE 7-31

Proper positioning for lateral oblique temporomandibular joint projection.

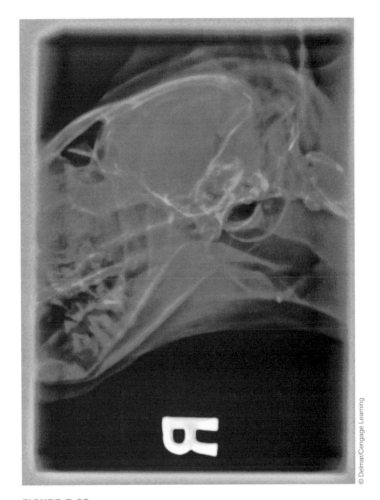

FIGURE 7-32

Lateral oblique temporomandibular joint projection.

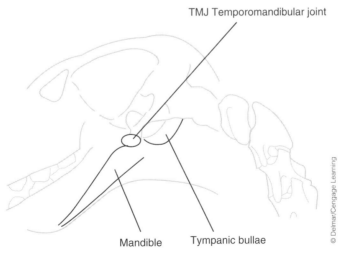

FIGURE 7-33

Anatomical features and landmarks: temporomandibular joint, tympanic bulla, and mandible.

DV Temporomandibular Joint Projection

Positioning:

- Sternal recumbency.
- Sandbag is placed across the cervical region to maintain placement of the head against the x-ray cassette.
- Tape can be used across the maxilla to maintain vertical alignment of the head on the x-ray cassette.

Centering:

- Palpate the base of the ear to determine placement just rostral on the skull. Center the beam on the dorsal midline of the skull just rostral to the bullae or ears.

Collimation:

- Collimate just cranial and caudal to the joint.
- Right and left side of skull fully within collimated area.

Labeling:

- Mark either left or right dependent side, or both.

Technique:

- Measure at the widest point of the cranium just caudal to the orbit and over the joint.

Comments:

- DV is preferred because the temporomandibular joint is located on the ventral portion of the skull and closer to the cassette.

© Delmar/Cengage Learning

FIGURE 7-34

Proper positioning for DV temporomandibular joint projection.

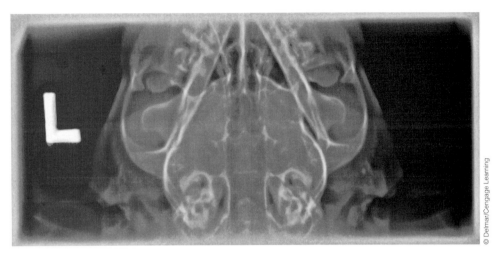

FIGURE 7-35

DV temporomandibular joint projection.

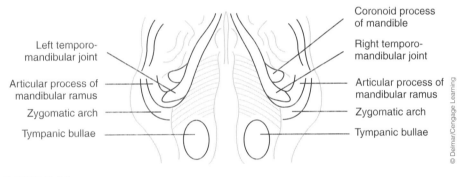

FIGURE 7-36

Anatomical features and landmarks: tympanic bulla, zygomatic arch, articular process of mandibular ramus, and coronoid process of mandible.

CHAPTER 8

DENTAL RADIOGRAPHS

OVERVIEW

Proper evaluation of the teeth requires specific positioning techniques to obtain an accurate and sufficiently detailed view of the teeth. Skull radiographs are not of sufficient detail to be diagnostic when evaluating oral pathology. Although standard x-ray machines can be used to position patients for dental radiology, exposures can be most easily accomplished with a radiology unit specific for dentistry.

Because the palate of dogs and cats is not arched as it is in humans, a bisecting angle technique is used to position the head of the x-ray unit to create an exact image of the tooth on the film. Because the roots are embedded in the maxillary bone, if the film was parallel to the tooth in the maxillary arcade, the image of the roots would not be on the film because the palate would get in the way. To aid in understanding the bisecting angle technique, imagine a person standing in a flat desert with nothing around. The sun (beam) is directly over the individual's head. A shadow (image) would be a very undistinguishable blob on the ground (film). It would not reveal any important information as to the shape and size of the individual. Alternatively, if the sun (beam) sat just above the horizon, the shadow would be 30 feet long, and still would not give any accurate information as to shape and size of the individual. If the sun was positioned exactly halfway between its location at high noon and just before sunset, the shadow behind the individual would be as tall as the individual and would represent accurate details about the individual's image. At this exact point, the sun is creating a beam that is hitting the bisecting angle perpendicularly.

Intraoral film is positioned so that the vertical axis of the tooth and the horizontal axis of the film create an angle. Bisect that angle or cut that angle in half with an imaginary line. The x-ray beam is then placed exactly perpendicular to that bisecting plane. If the angle of the tube is too steep, the image will be foreshortened. If the angle of the tube is too shallow, the image will be too elongated.

The following pages illustrate the proper positioning for dental radiographs. Note that there are slight differences in the views taken for dogs and cats, so each species is described separately. Also note that collimation for intraoral x-ray units is automatic, and film identification labels are always placed on the dimple convex side of the film toward the x-ray tube so each required position contains descriptive information for the patient positioning and centering of the x-ray beam.

FIGURE 8-1

Proper positioning for canine upper incisor arcade.

Canine Upper Incisor Arcade

Positioning:

- Lateral or dorsal recumbency.

Centering:

- Center image so that two middle incisors are equidistant from edge of image and visualize at least 3 mm periapical bone.

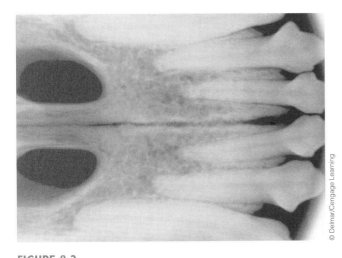

FIGURE 8-2

Canine upper incisor arcade.

FIGURE 8-3

Proper positioning for canine upper canine tooth.

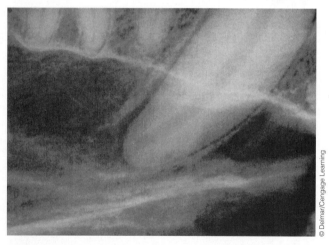

FIGURE 8-4

Canine upper canine tooth.

Canine Upper Canine Tooth

Positioning:

- Lateral or dorsal recumbency.

Centering:

- Apex of tooth visible with at least 3 mm periapical bone.

Canine Upper Premolars

Positioning:

- Lateral or dorsal recumbency.

Centering:

- Apex of teeth visible with at least 3 mm periapical bone.

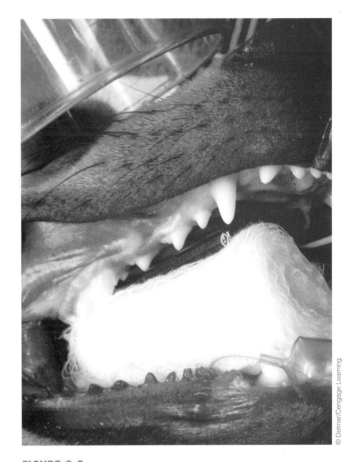

FIGURE 8-5

Proper positioning for canine upper premolars.

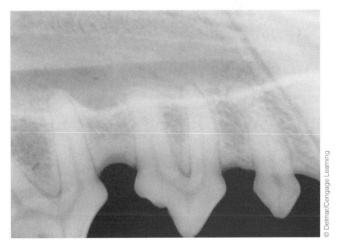

FIGURE 8-6

Canine upper premolars.

Canine Upper Fourth Premolars and Molars

Positioning:

- Lateral or dorsal recumbency.

Centering:

- Apex of teeth visible with at least 3 mm periapical bone.
- Swing tube head rostrally or distally to separate superimposed mesial buccal and palatal roots of fourth premolar.

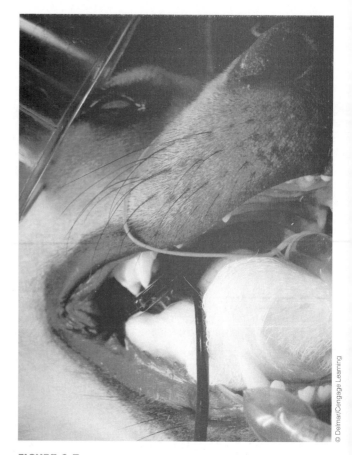

FIGURE 8-7

Proper positioning for canine upper fourth premolars and molars.

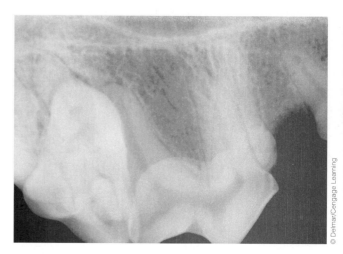

FIGURE 8-8

Canine upper fourth premolars and molars.

FIGURE 8-9

Proper positioning for canine lower incisor arcade.

Canine Lower Incisor Arcade

Positioning:

* Lateral or dorsal recumbency.

Centering:

* Center image so that two middle incisors are equidistant from edge of image and visualize at least 3 mm periapical bone.

FIGURE 8-10

Canine lower incisor arcade.

Canine Lower Canine Tooth and Lower Premolar Arcade

Positioning:

- Lateral or dorsal recumbency.

Centering:

- Apex of tooth visible with at least 3 mm periapical bone.

Comments:

- View may be taken with bisecting angle or parallel technique.

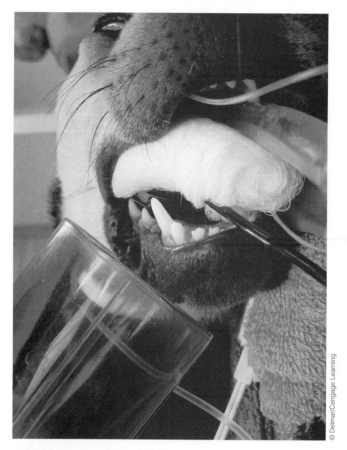

FIGURE 8-11

Proper positioning for canine lower canine tooth and lower premolar arcade.

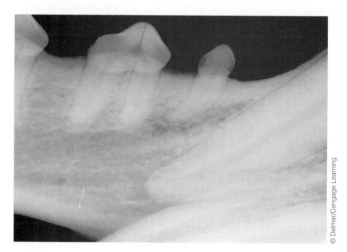

FIGURE 8-12

Canine lower canine tooth and lower premolar arcade.

Canine Lower Premolar Arcade

Positioning:

- Lateral or dorsal recumbency.

Centering:

- Apex of teeth visible with at least 3 mm periapical bone.

Comments:

- View is to be taken using parallel technique.

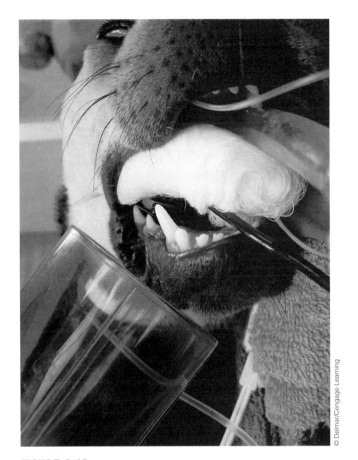

FIGURE 8-13

Proper positioning for canine lower premolar arcade.

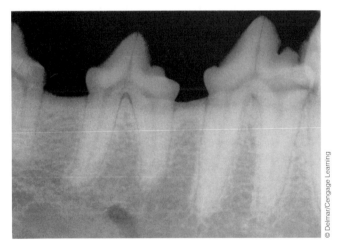

FIGURE 8-14

Canine lower premolar arcade.

FIGURE 8-15

Proper positioning for canine lower molars.

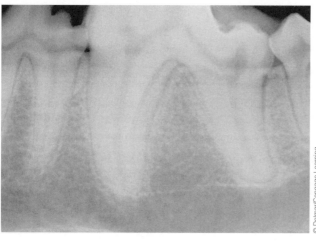

FIGURE 8-16

Canine lower molars.

Canine Lower Molars

Positioning:

* Lateral or dorsal recumbency.

Centering:

* Apex of teeth visible with at least 3 mm periapical bone.

Comments:

* View is to be taken using parallel technique.

FIGURE 8-17

Proper positioning for feline upper incisor arcade.

Feline Upper Incisor Arcade

Positioning:

- Lateral or dorsal recumbency.

Centering:

- Center image so that two middle incisors are equidistant from edge of image and visualize at least 3 mm periapical bone.

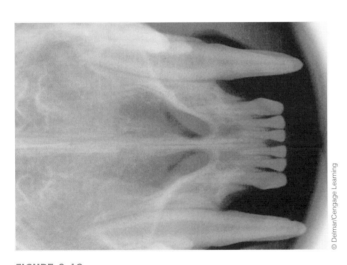

FIGURE 8-18

Feline upper incisor arcade.

Feline Upper Canine Tooth

Positioning:

- Lateral or dorsal recumbency.

Centering:

- Apex of tooth visible with at least 3 mm periapical bone.

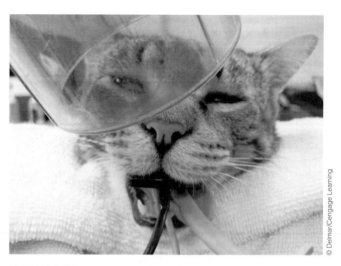

FIGURE 8-19

Proper positioning for feline upper canine tooth.

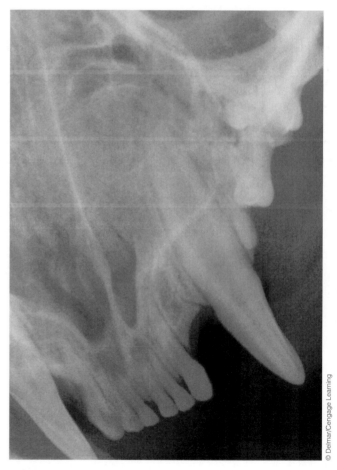

FIGURE 8-20

Feline upper canine tooth.

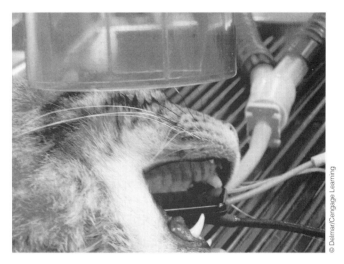

FIGURE 8-21

Proper positioning for feline upper premolars and molar.

Feline Upper Premolars and Molar

Positioning:

- Lateral or dorsal recumbency.

Centering:

- Apex of teeth visible with at least 3 mm periapical bone.
- Swing tube head rostrally or distally to separate superimposed mesial buccal and palatal roots of fourth premolar.

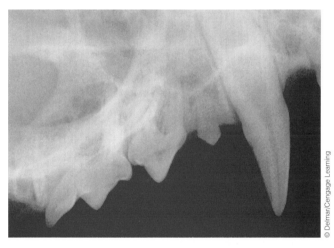

FIGURE 8-22

Feline upper premolars and molar.

Feline Lower Incisor Arcade

Positioning:

- Lateral or dorsal recumbency.

Centering:

- Center image so that two middle incisors are equidistant from edge of image and visualize at least 3 mm periapical bone.

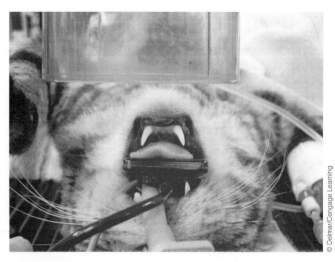

FIGURE 8-23

Proper positioning for feline lower incisor arcade.

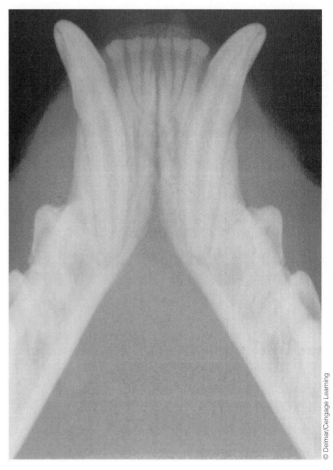

FIGURE 8-24

Feline lower incisor arcade.

Feline Lower Canine Tooth

Positioning:

- Lateral or dorsal recumbency.

Centering:

- Apex of tooth visible with at least 3 mm periapical bone.

FIGURE 8-25

Proper positioning for feline lower canine tooth.

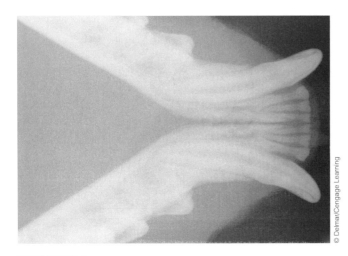

FIGURE 8-26

Feline lower canine tooth.

FIGURE 8-27

Proper positioning for feline lower premolars and molar arcade.

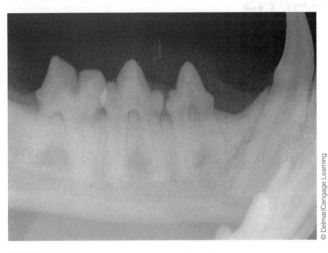

FIGURE 8-28

Feline lower premolars and molar arcade.

Feline Lower Premolars and Molar Arcade

Positioning:

- Lateral or dorsal recumbency.

Centering:

- Apex of teeth visible with at least 3 mm periapical bone.

CHAPTER 9

SPINAL RADIOGRAPHS

OVERVIEW

Radiographs of the vertebral column are used to detect bony lesions as well as evaluate intervertebral disc space. Careful positioning is necessary to maintain the vertebral column parallel to the x-ray cassette and to allow the vertebral column to be placed as near to the x-ray cassette as possible. Positioning aids are used as supportive devices to maintain the spine parallel to the tabletop. Placing a piece of tape along the spinal column before moving the patient onto the x-ray table may aid in maintaining proper alignment (Figure 9-1).

Commonly obtained views of the vertebral column include lateral and ventrodorsal (VD) projections of the cervical area, thoracic area, thoracolumbar junction, lumbar area, lumbosacral area, sacral area, and coccygeal (caudal) vertebrae. Cervical spine evaluations may also require flexed and extended views. To enhance detail, spinal radiographs are tightly collimated. Care must be taken to ensure that directional and identification labels placed before making the exposure are within the collimated area and not overlapping any portion of the vertebral column. The following pages illustrate the proper positioning and technique for spinal radiographs.

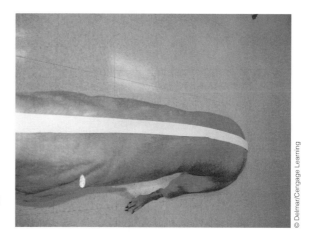

FIGURE 9-1

Use of tape to maintain proper alignment.

151

VD Cervical Spine Projection

Positioning:

- Dorsal recumbency.
- Foam pad is placed under neck to maintain spinal column parallel to x-ray cassette.
- Forelimbs are secured evenly and caudally.

Centering:

- C4–C5 intervertebral space.

Collimation:

- Base of the skull to the spine of the scapula.

Labeling:

- R/L marker within collimated area away from bony areas.
- Identification label in caudal region.

Technique:

- Measure at the area of the C4–C5 intervertebral space.

Comments:

- For very large patients, when there is a significant difference in measurement between the cranial and caudal areas of the cervical spine, two views should be taken. The first view is centered on the C2–C3 space and collimated from the base of the skull to C4. The second view is remeasured and centered on the C5–C6 space and collimated to contain C4–T1.

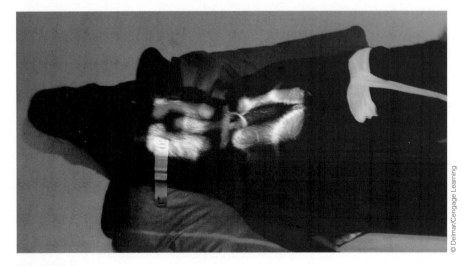

FIGURE 9-2

Proper position for VD cervical spine projection.

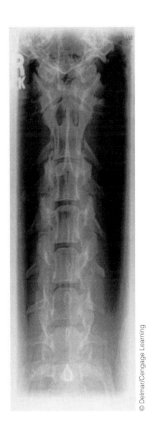

FIGURE 9-3

VD cervical spine projection.

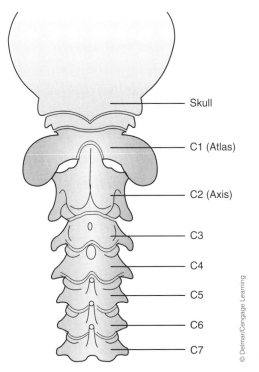

Skull

C1 (Atlas)

C2 (Axis)

C3

C4

C5

C6

C7

FIGURE 9-4

Anatomical features and landmarks: skull, C1 (atlas), C2 (axis), C3, C4, C5, C6, and C7.

Lateral Cervical Spine Projection

Positioning:

- Right or left lateral recumbency.
- Foam pad is placed under mandible to maintain spinal column parallel to x-ray cassette and secured with a sandbag.
- Forelimbs are secured evenly and caudally.
- Foam pad may be needed along the sternum to avoid rotation of the spinal column.

Centering:

- C4–C5 intervertebral space.

Collimation:

- Base of the skull to the spine of the scapula.

Labeling:

- R/L marker within collimated area away from bony areas to indicate side facing closest to x-ray cassette.
- Identification label in caudal region.

Technique:

- Measure at the area of the C4–C5 intervertebral space.

Comments:

- Neck should be in a natural position, not flexed or extended.

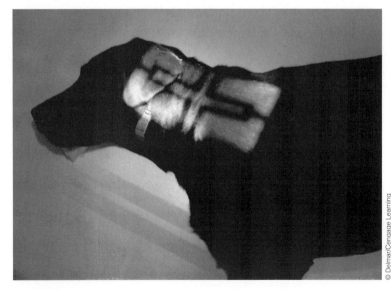

© Delmar/Cengage Learning

FIGURE 9-5

Proper position for lateral cervical spine projection.

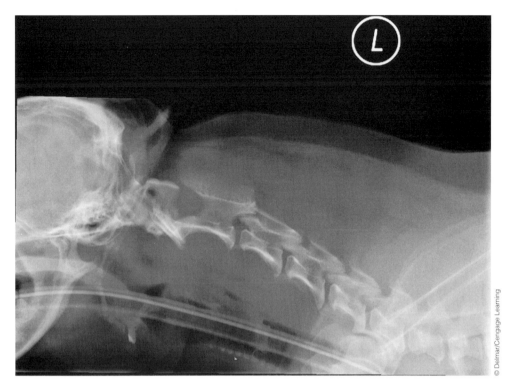

FIGURE 9-6

Lateral cervical spine projection.

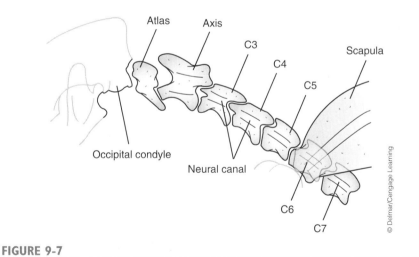

FIGURE 9-7

Anatomical features and landmarks: occipital condyle, atlas, axis, C3, C4, neural canal, C5, and scapula.

Lateral Cervical Spine Extended Projection

Positioning:

- Right or left lateral recumbency.
- Foam pad is placed under mandible to maintain spinal column parallel to x-ray cassette and secured with a sandbag.
- Forelimbs are secured evenly and caudally.
- Foam pad may be needed along the sternum to avoid rotation of the spinal column.
- The neck is extended or pushed dorsally.

Centering:

- C4–C5 intervertebral space.

Collimation:

- Base of the skull to the spine of the scapula.

Labeling:

- R/L marker within collimated area away from bony areas to indicate side facing closest to x-ray cassette.
- Identification label in caudal region.

Technique:

- Measure at the area of the C4–C5 intervertebral space.

Comments:

- To be in true extension, it is not enough just to push the skull dorsally; the whole cervical spine from C7 cranial needs to be extended dorsally.

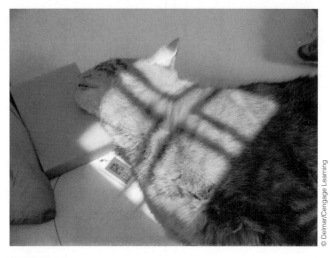

FIGURE 9-8

Proper position for lateral cervical spine extended projection.

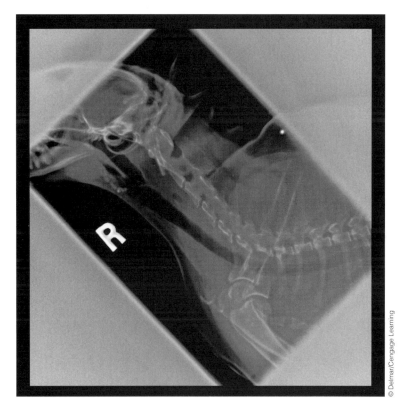

FIGURE 9-9

Lateral cervical spine extended projection.

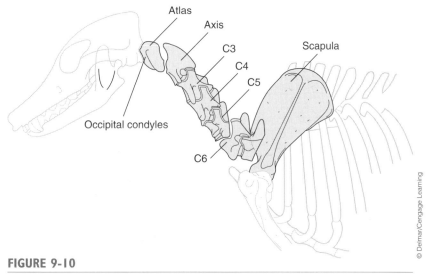

FIGURE 9-10

Anatomical features and landmarks: occipital condyles, atlas, axis, C3, C4, C5, scapula, and C6.

Lateral Cervical Spine Flexed Projection

Positioning:

- Right or left lateral recumbency.
- Head is directed ventrally and caudally toward the humeri and can be secured with a sandbag to maintain flexion on the dorsal part of the skull.
- Forelimbs are secured evenly and caudally.
- Foam pad may be needed along the sternum to avoid rotation of the spinal column.

Centering:

- C4–C5 intervertebral space.

Collimation:

- Base of the skull to the spine of the scapula.

Labeling:

- R/L marker within collimated area away from bony areas to indicate side facing closest to x-ray cassette.
- Identification label in caudal region.

Technique:

- Measure at the area of the C4–C5 intervertebral space.

Comments:

- Take care not to hyperflex the neck. Flexion must be even throughout all cervical vertebrae starting at C7.

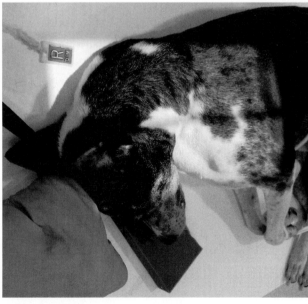

FIGURE 9-11

Proper position for lateral cervical spine flexed projection.

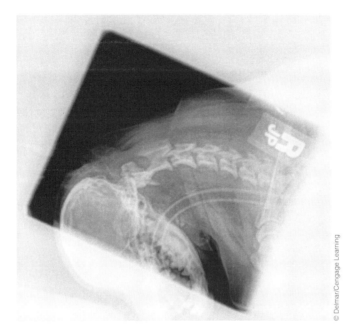

FIGURE 9-12

Lateral cervical spine flexed projection.

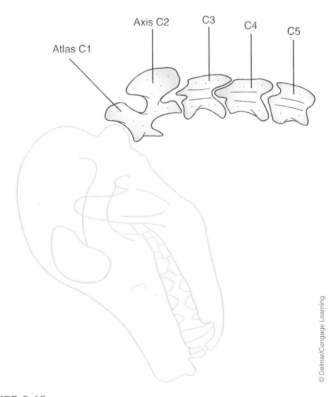

FIGURE 9-13

Anatomical features and landmarks: atlas C1, axis C2, C3, C4, and C5.

VD Thoracic Spine Projection

Positioning:

- Dorsal recumbency.
- Forelimbs extended evenly and cranially.
- V-trough or sandbags to maintain vertical alignment.

Centering:

- Caudal border of scapula at approximately the sixth or seventh thoracic vertebra.

Collimation:

- Halfway between xiphoid and last rib to spine of the scapula.
- Must include C7–L1.

Labeling:

- R/L marker within collimated area away from bony areas.
- Identification label in caudal region.

Technique:

- Measure at highest (thickest) point of sternum.

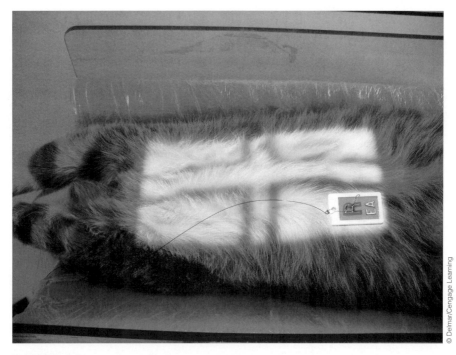

FIGURE 9-14

Proper position for VD thoracic spine projection.

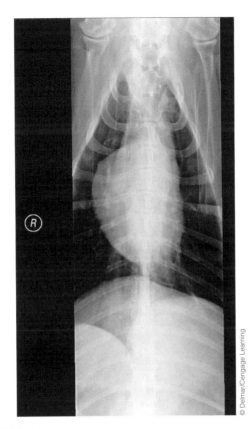

FIGURE 9-15

VD thoracic spine projection.

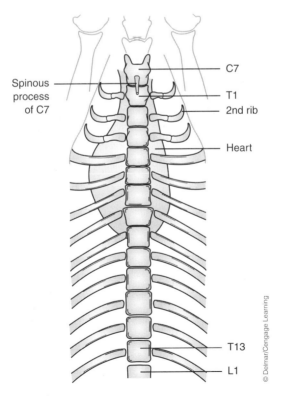

FIGURE 9-16

Anatomical features and landmarks: L1, T13, T1, C7, spinous process of C7, 2nd rib, and heart.

Lateral Thoracic Spine Projection

Positioning:

- Right or left lateral recumbency.
- Forelimbs are extended evenly and slightly cranially.
- Hindlimbs are extended evenly and slightly caudally.
- Foam pad may be needed along the sternum to avoid rotation of the spinal column.

Centering:

- Caudal border of scapula at approximately the sixth or seventh thoracic vertebra.

Collimation:

- Halfway between xiphoid and last rib to spine of the scapula.
- Must include C7–L1.

Labeling:

- R/L marker within collimated area away from bony areas to indicate side facing closest to x-ray cassette.
- Identification label in caudal region.

Technique:

- Measure at the midpoint of the xiphoid or the highest point of the thorax.

© Delmar/Cengage Learning

FIGURE 9-17

Proper position for lateral thoracic spine projection.

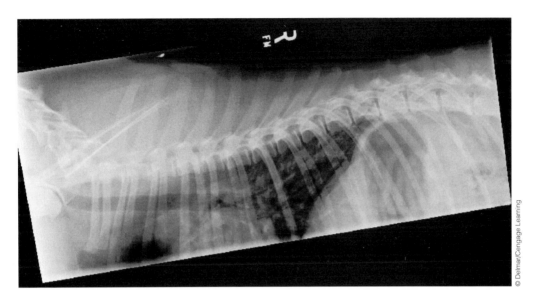

FIGURE 9-18

Lateral thoracic spine projection.

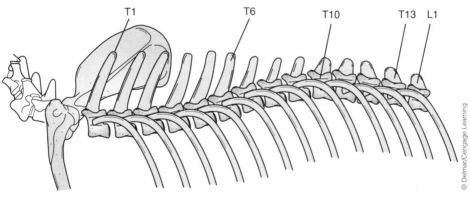

FIGURE 9-19

Anatomical features and landmarks: T1, T6, T10, T12, T13, and L1.

VD Thoracolumbar Spine Projection

Positioning:

- Dorsal recumbency.
- Forelimbs are extended evenly and slightly cranially.
- Hindlimbs are extended evenly and slightly caudally.
- Foam pad may be needed along the sternum to avoid rotation of the spinal column.

Centering:

- Halfway between collimation points.

Collimation:

- Xiphoid to last rib.

Labeling:

- R/L marker within collimated area away from bony areas.
- Identification label in caudal region.

Technique:

- Measure at the midpoint of the xiphoid or the highest point of the thorax.

FIGURE 9-20

Proper position for VD thoracolumbar spine projection.

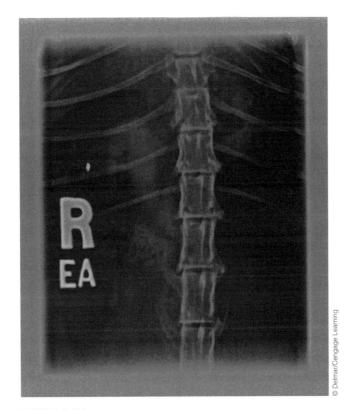

FIGURE 9-21

VD thoracolumbar spine projection.

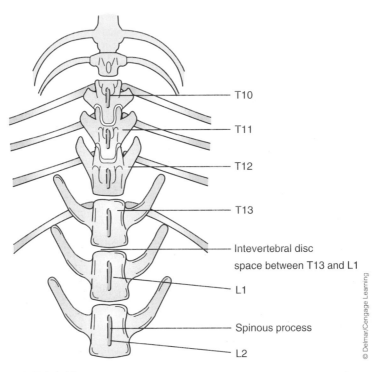

T10

T11

T12

T13

Intevertebral disc space between T13 and L1

L1

Spinous process

L2

FIGURE 9-22

Anatomical features and landmarks: T10–L2, intervertebral disc space between T13 and L1, and spinous process.

Lateral Thoracolumbar Spine Projection

Positioning:

- Right or left lateral recumbency.
- Forelimbs are extended evenly and slightly cranially.
- Hindlimbs are extended evenly and slightly caudally.
- Foam pad may be needed along the sternum to avoid rotation of the spinal column.

Centering:

- Halfway between collimation points.

Collimation:

- Xiphoid to last rib.

Labeling:

- R/L marker within collimated area away from bony areas to indicate side facing closest to x-ray cassette.

Technique:

- Measure at the midpoint of the xiphoid or the highest point of the thorax.

© Delmar/Cengage Learning

FIGURE 9-23

Proper position for lateral thoracolumbar spine projection.

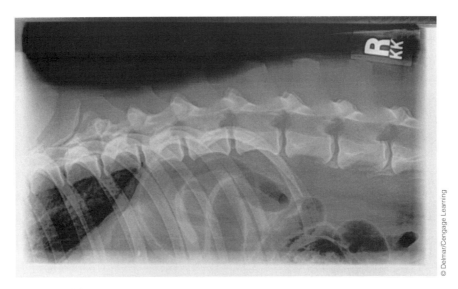

FIGURE 9-24

Lateral thoracolumbar spine projection.

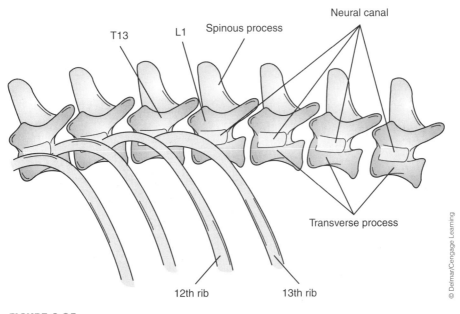

FIGURE 9-25

Anatomical features and landmarks: T13, L1, 12th rib, 13th rib, neural canal, spinous process, and transverse process.

VD Lumbar Spine Projection

Positioning:

- Dorsal recumbency.
- Forelimbs are extended evenly and slightly cranially.
- Hindlimbs are extended evenly and slightly caudally.
- Foam pad may be needed along the sternum to avoid rotation of the spinal column.

Centering:

- Palpate xiphoid and the wing of the ilium, and place the center halfway between these two points.

Collimation:

- Xiphoid to acetabulum.

Labeling:

- R/L marker within collimated area away from bony areas.
- Identification label in caudal region.

Technique:

- Measure at mid-lumbar spine.

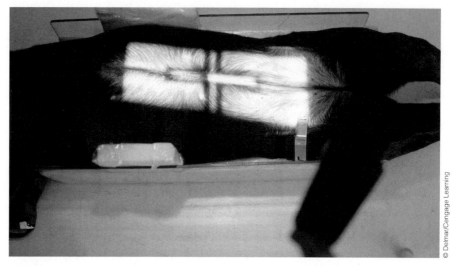

© Delmar/Cengage Learning

FIGURE 9-26

Proper position for VD lumbar spine projection.

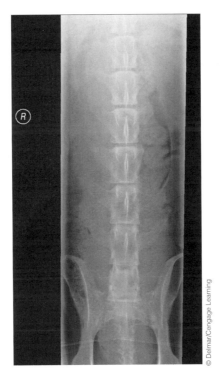

FIGURE 9-27

VD lumbar spine projection.

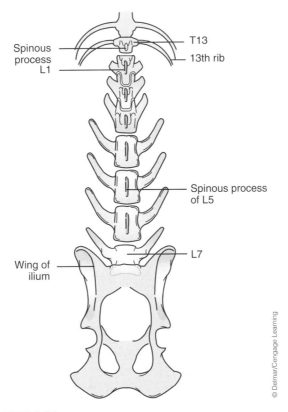

FIGURE 9-28

Anatomical features and landmarks: 13th rib, spinous process, T13, L1, spinous process of L5, L7, and wing of ilium.

Lateral Lumbar Spine Projection

Positioning:

- Right or left lateral recumbency.
- Forelimbs are extended evenly and slightly cranially.
- Hindlimbs are extended evenly and slightly caudally.
- Foam pad may be needed along the sternum to avoid rotation of the spinal column.

Centering:

- L3–4.

Collimation:

- Xiphoid to acetabulum.

Labeling:

- R/L marker within collimated area away from bony areas to indicate side facing closest to x-ray cassette.

Technique:

- Measure mid-lumbar spine.

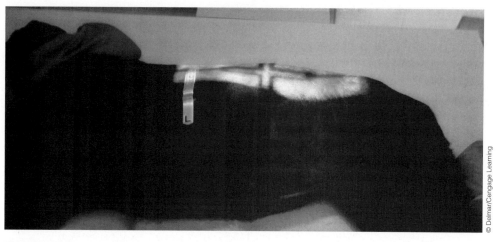

FIGURE 9-29

Proper position for lateral lumbar spine projection.

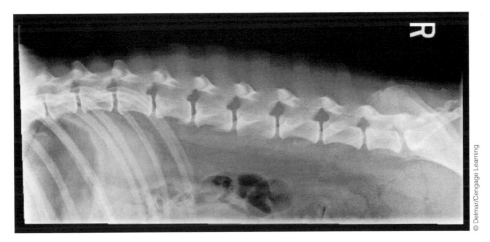

FIGURE 9-30

Lateral lumbar spine projection.

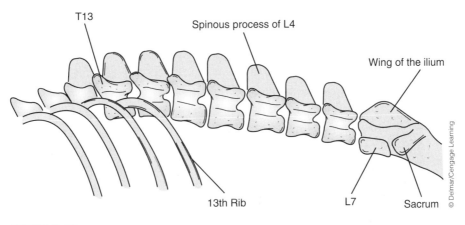

FIGURE 9-31

Anatomical features and landmarks: T13, 13th rib, spinous process of L4, wing of the ilium, L7, and sacrum.

VD Lumbosacral Spine Projection

Positioning:

- Dorsal recumbency in V-trough.
- Forelimbs are extended evenly and slightly cranially.
- Hindlimbs are extended evenly and slightly caudally.
- Foam pad may be needed along the sternum to avoid rotation of the spinal column.

Centering:

- Palpate the wing of the ilium, and center just caudal and midpelvis.

Collimation:

- Sixth lumbar vertebra to iliac crest.

Labeling:

- R/L marker within collimated area away from bony areas.
- Identification label in caudal region.

Technique:

- Measure at the wing of the ilium.

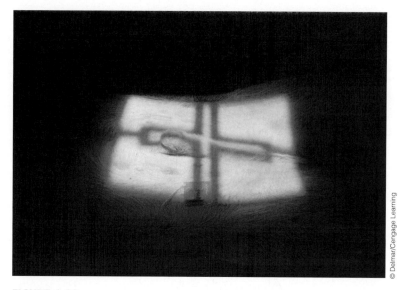

© Delmar/Cengage Learning

FIGURE 9-32

Proper position for VD lumbosacral spine projection.

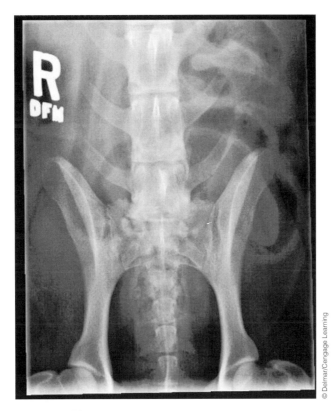

FIGURE 9-33

VD lumbosacral spine projection.

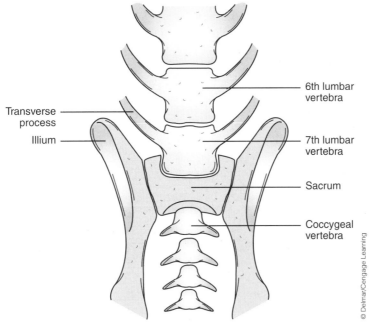

FIGURE 9-34

Anatomical features and landmarks: 7th lumbar vertebra, sacrum, coccygeal vertebra, ilium, 6th lumbar vertebra, and transverse process.

Lateral Lumbosacral Spine Projection

Positioning:

- Right or left lateral recumbency.
- Forelimbs are extended evenly and slightly cranially.
- Hindlimbs are extended evenly and slightly caudally.
- Foam pad may be needed along the sternum to avoid rotation of the spinal column.
- Foam wedge is placed between hindlimbs to superimpose both sides of pelvis.

Centering:

- Palpate the wing of the ilium and the dorsal spinous process of the lumbosacral region, and center just caudal to the wing of the ilium. There is a distinct divot at the lumbosacral junction.

Collimation:

- Cranial to include L-6 lumbar vertebrae to iliac crest to include the first cranial caudal vertebrae (tail).

Labeling:

- R/L marker is within collimated area away from bony areas to indicate side facing closest to x-ray cassette.
- Place identification label in right cranial region or left caudal region within collimated area to avoid overlap of bone.

Technique:

- Measure thickest area just caudal to the wing of the ilium.

Comments:

- Alternate views include flexed and extended views of the lumbosacral junction.

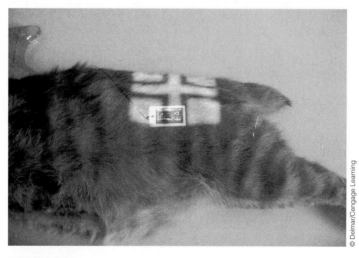

FIGURE 9-35

Proper position for lateral lumbosacral spine projection.

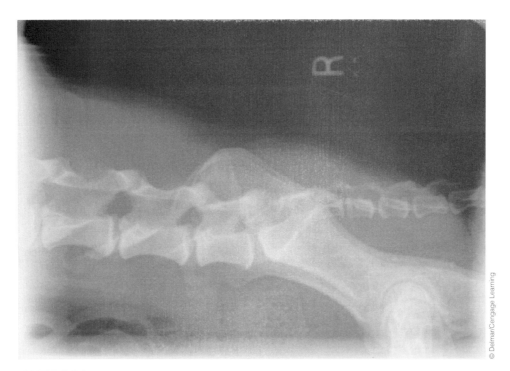

FIGURE 9-36

Lateral lumbosacral spine projection.

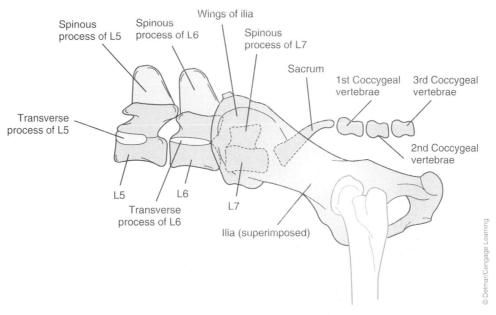

FIGURE 9-37

Anatomical features and landmarks: 1st coccygeal vertebrae, 2nd coccygeal vertebrae, 3rd coccygeal vertebrae, sacrum, spinous process of L7, spinous process of L6, spinous process of L5, transverse process of L5, transverse process of L6, L5, L6, L7, ilia (superimposed), and wings of ilia.

VD Coccygeal (Caudal) Spine Projection

Positioning:

- Dorsal recumbency.
- V-trough or sandbags to maintain dorsal recumbency.
- Hindlimbs in natural position.
- Tail extended caudally.

Centering:

- Midway from the sacrum to the tip of the tail.

Collimation:

- Cranial to the sacrum to the tip of the tail.

Labeling:

- R/L marker within collimated area away from bony areas.
- Identification in caudal region.

Technique:

- Measure at thickest part of tail.

Comments:

- Tape may be used to maintain vertical alignment of the tail.

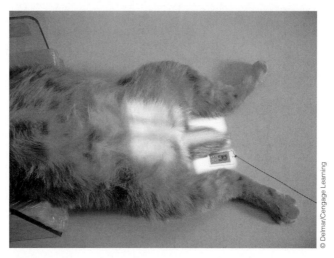

FIGURE 9-38

Proper position for VD coccygeal (caudal) spine projection.

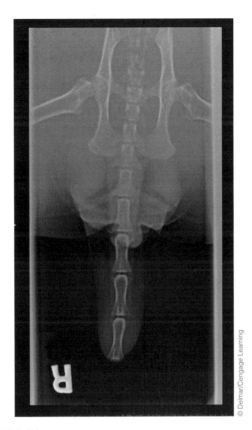

FIGURE 9-39

VD coccygeal (caudal) spine projection.

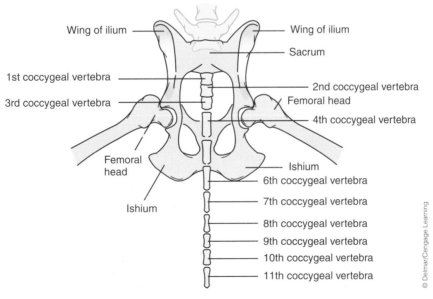

FIGURE 9-40

Anatomical features and landmarks: sacrum, wing of ilium, 1st coccygeal vertebra, 3rd coccygeal vertebra, 2nd coccygeal vertebra, ischium, femoral head, 4th coccygeal vertebra, 6th coccygeal vertebra, 7th coccygeal vertebra, 8th coccygeal vertebra, 9th coccygeal vertebra, 10th coccygeal vertebra, and 11th coccygeal vertebra.

Lateral Coccygeal (Caudal) Spine Projection

Positioning:

- Right or left lateral recumbency.

Centering:

- Midway from the sacrum to the tip of the tail.

Collimation:

- Cranial to the sacrum to the tip of the tail.

Labeling:

- R/L marker within collimated area away from bony areas.
- Identification label in caudal region.

Technique:

- Measure at thickest part of tail.

Comments:

- Cassette may be elevated off the table to maintain alignment of the tail and spine, and bring the tail closer to the x-ray cassette, or a sponge may be placed under the tail to make it parallel to the cassette; this will depend on the size of the patient
- Tape may be needed to keep the tail in vertical alignment.

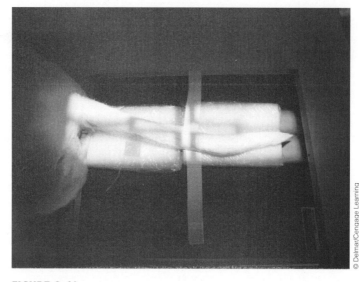

© Delmar/Cengage Learning

FIGURE 9-41

Proper position for lateral coccygeal (caudal) spine projection.

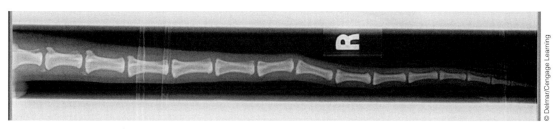

FIGURE 9-42

Lateral coccygeal (caudal) spine projection.

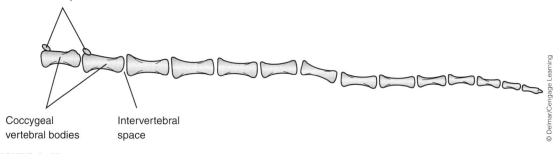

FIGURE 9-43

Anatomical features and landmarks: cranial articular processes, coccygeal vertebral bodies, and intervertebral space.

CHAPTER 10

RADIOGRAPHY OF AVIAN AND EXOTIC ANIMALS

OVERVIEW

There are many unique aspects of radiology relating to obtaining diagnostic quality radiographs of birds, reptiles, amphibians, and small mammals. The type of film and cassettes used, the focal point, collimation, exposure time, and restraint procedures require modification from the standard techniques used for dogs and cats.

Mammography film and cassettes are routinely used for avian and exotic animal radiology. Mammography film is a single emulsion, and the cassettes are single intensifying screen cassettes. The film is placed in the cassette with the dark side of the film facing the dark side of the cassette, and the light side facing the light side (Figure 10-1).

The cassette is placed on the tabletop and the x-ray tube head moved to achieve a focal-spot distance of 40 inches (100 cm). This results in a magnification of the image on the film and aids in evaluation of the radiograph in these small species. Close collimation around the area of interest will help decrease scatter radiation and maximize detail.

Exposure time and settings used will vary with the type of film and radiographic equipment available. When using mammography film, a longer exposure time is required. A milliamperage (mA) setting of 300, exposure time of 1/10 second, and a kilovolt peak (kVp) between 40 and 50 will generally yield a high-quality, high-detail image. When using double-sided film, an mA setting of 300, exposure time of 1/60 second, and a kVp between 40 and 50 will achieve a good-quality film.

GENERAL PRINCIPLES OF RESTRAINT

Proper patient restraint is required to minimize or eliminate any movement that will result in a blurred or low-quality image. This can be accomplished with either physical restraint or chemical restraint.

In many avian patients, physical restraint is generally considered unacceptable, especially in large, powerful birds, highly stressed or fractious birds, and birds with injuries that can be exacerbated with struggling. In these cases, chemical restraint is used. Inhalation anesthesia (sevoflurane or isoflurane) is suitable for chemical restraint. In some calm birds, physical restraint using tape or commercial restraint devices can be used. Heavy metal examination of avian patients can be quickly accomplished by placing a small bird in a paper bag (Figure 10-2). The bag is then placed directly on the x-ray cassette. Large birds can stand directly on the cassette, or the horizontal beam can be used and the bird placed in a perch.

Most lizards and turtles will quietly sit on the cassette without any restraint for the dorsal-ventral view. More excitable reptiles can be restrained with the use of tape, sandbags, or cotton ball Vetwrap hoods (Figures 10-3 and 10-4). The cotton ball Vetwrap hood helps

FIGURE 10-1

Proper film placement in cassette.

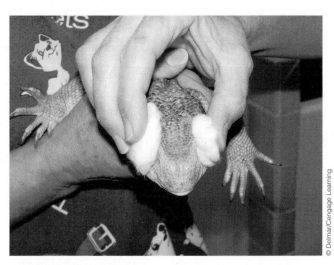

FIGURE 10-3

Restraint techniques for reptiles.

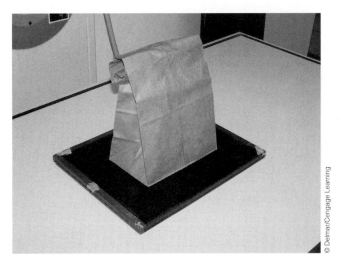

FIGURE 10-2

Radiography of birds.

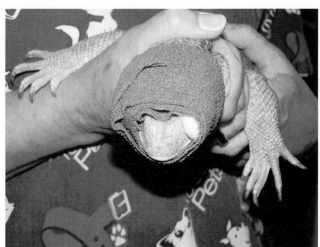

FIGURE 10-4

Restraint technique for reptiles.

calm the patient. Very fractious or aggressive patients may need some form of chemical restraint.

Restraint of a snake can be challenging. Tape and sandbag restraints prove to be ineffective. Generally, manual restraint is the best. Manual restraint or the use of acrylic tubes is the most common form of restraint for snakes (Figures 10-5 and 10-6).

Rabbits and ferrets can be physically restrained with techniques similar to those used with cats. Anesthesia may be recommended in the rabbit or ferret that is fractious or stressed.

FIGURE 10-5

Restraint of snakes.

FIGURE 10-6

Radiography of snakes.

Lateral View of the Avian Patient

Positioning:

- Right lateral recumbency.
- Neck is extended, and sponges are used to support the head, parallel to the cassette.
- Wings are extended dorsally and secured with tape at the carpal joint.
- Sternum should be parallel to the cassette.
- Legs are pulled caudally and secured with tape.

Centering:

- Midsternum.

Collimation:

- Whole body.

Labeling:

- Right marker within the collimated area in cranial aspect of the sternum.

Technique:

- Measure the thickest point of the chest.

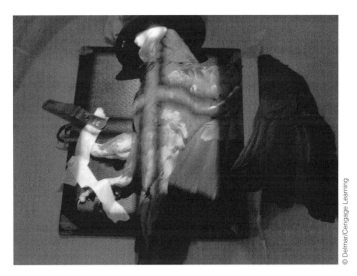

FIGURE 10-7

Proper positioning for lateral view of the avian patient.

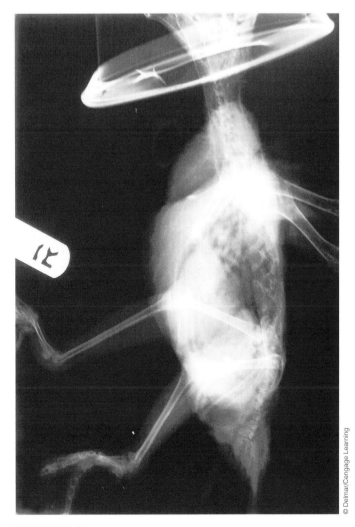

FIGURE 10-8

Lateral view of the avian patient.

Ventrodorsal View of the Avian Patient

Positioning:

- Dorsal recumbency.
- Neck is extended, and sponges are used to support the head, parallel to the cassette.
- Wings are extended lateral from the body and secured with tape at the carpal joint.
- Sternum should be placed directly over the spinal column.
- Legs are pulled caudally and secured individually with tape.

Centering:

- Midsternum.

Collimation:

- Whole body.

Labeling:

- R/L marker within the collimated area.

Technique:

- Measure the thickest point of the chest.

FIGURE 10-9

Proper positioning for ventrodorsal view of the avian patient.

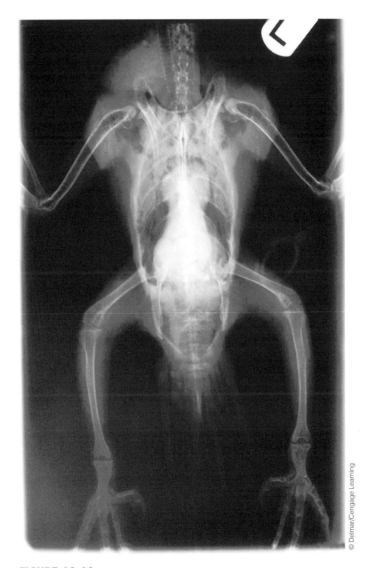

FIGURE 10-10

Ventrodorsal view of the avian patient.

Dorsoventral View of the Lizard

Positioning:

- Ventral recumbency.
- Forelimbs and hindlimbs are gently placed lateral to the body.
- Tape may be used to secure in position.

Centering:

- Midbody region.

Collimation:

- Whole body, include the head, legs, and cranial aspect of the tail.

Labeling:

- R/L marker within collimated area.

Technique:

- Measure the thickest point of the body.

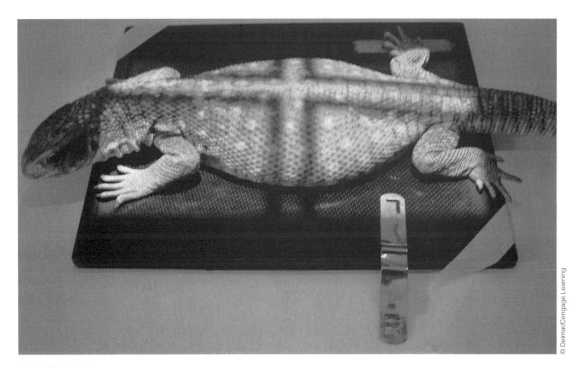

FIGURE 10-11

Proper positioning for dorsoventral view of the lizard.

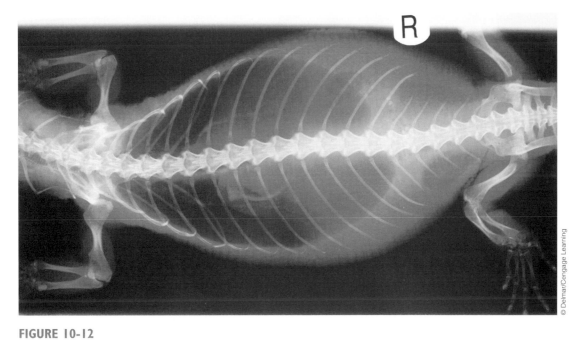

FIGURE 10-12

Dorsoventral view of the lizard.

Lateral View of the Lizard

Positioning:

- Right lateral recumbency.
- Forelimbs are pulled ventral and cranial of the body.
- Hindlimbs are pulled ventral and caudal of the body.
- Both forelimbs and hindlimbs can be secured with Vetwrap to minimize struggling (Figure 10-13).
- Secure limbs with tape.
- Tape across shoulders and hips to secure the body; sponges may be used to maintain horizontal position.

Centering:

- Midbody region.

Collimation:

- Whole body, include the head, legs, and the cranial aspect of the tail.

Labeling:

- R/L marker within collimated area.

Technique:

- Measure the thickest point of the body.

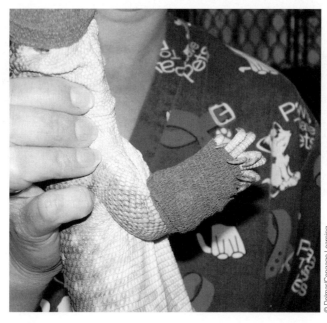

FIGURE 10-13

Restraint of the lizard.

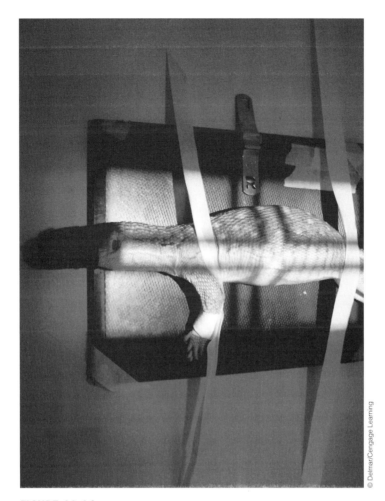

FIGURE 10-14

Proper positioning for lateral view of the lizard.

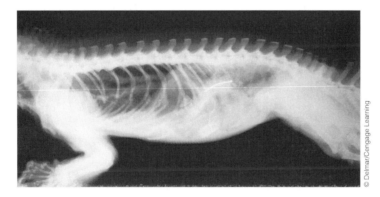

FIGURE 10-15

Lateral view of the lizard.

Lateral View of the Lizard with the Horizontal Beam

Positioning:

- Ventrodorsal recumbency.
- Ensure that the body is as close to the cassette as possible.

Centering:

- Midbody region.

Collimation:

- Whole body, including head and cranial aspect of the tail.

Labeling:

- R/L marker taped to the cassette above the patient.

Technique:

- Measure the thickest point of the body.

FIGURE 10-16

Proper positioning for lateral view of the lizard with the horizontal beam.

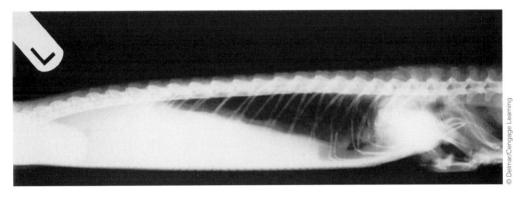

FIGURE 10-17

Lateral view of the lizard with the horizontal beam.

Dorsoventral View of the Turtle

Positioning:

- Ventral recumbency.
- Forelimbs and hindlimbs are gently placed lateral to the body.
- Tape may be used to secure in position.

Centering:

- Midbody region.

Collimation:

- Whole body, include the head, legs, and cranial aspect of the tail.

Labeling:

- R/L marker within collimated area.

Technique:

- Measure the thickest point of the body.

FIGURE 10-18

Proper positioning for dorsoventral view of the turtle.

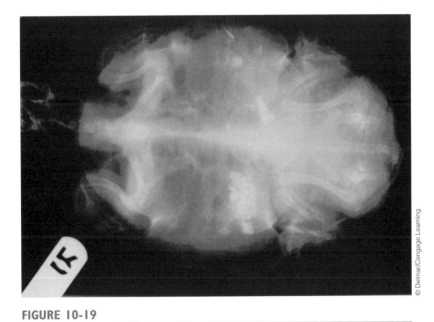

FIGURE 10-19

Dorsoventral view of the turtle.

Lateral View of the Turtle with the Horizontal Beam

Positioning:

- Ventrodorsal recumbency.
- Ensure that the body is as close to the cassette as possible.

Centering:

- Midbody region.

Collimation:

- Whole body, including head and cranial aspect of the tail.

Labeling:

- R/L marker taped to the cassette above the patient.

Technique:

- Measure the thickest point of the body.

FIGURE 10-20

Proper positioning for lateral view of the turtle with the horizontal beam.

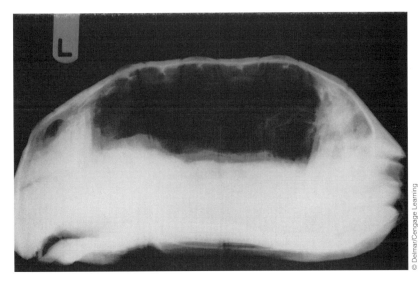

FIGURE 10-21

Lateral view of the turtle with the horizontal beam.

Lateral View of the Rabbit Abdomen

Positioning:

- Right lateral recumbency.
- Forelimbs are extended cranially and hindlimbs caudally; secure with tape.
- Sternum should be parallel to the cassette; this can be accomplished with the use of sponges.

Centering:

- Center slightly cranial of the last rib.

Collimation:

- Slightly cranial of the xiphoid and slightly caudal of the pubis.

Labeling:

- R marker is placed in the inguinal area.

Technique:

- Measure at the last rib.

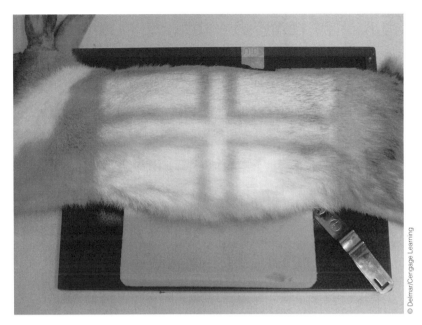

FIGURE 10-22

Proper positioning for lateral view of the rabbit abdomen.

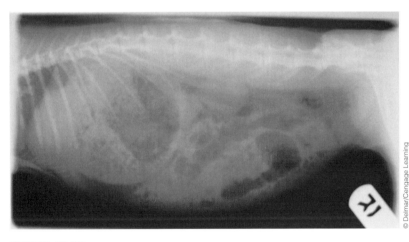

FIGURE 10-23

Lateral view of the rabbit abdomen.

Ventrodorsal View of the Rabbit Abdomen

Positioning:

- Dorsal recumbency.
- The front portion is secured with sandbags to keep the trunk bilaterally symmetrical.
- The hind limbs are extended and secured with sandbags or tape.

Centering:

- Slightly caudal to the last rib.

Collimation:

- Slightly cranial of the xiphoid and slightly caudal of the pubis.

Labeling:

- R marker is placed in the inguinal area.

Technique:

- Measure at the last rib.

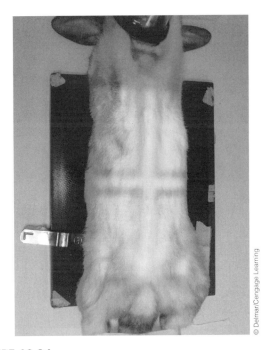

FIGURE 10-24

Proper positioning for ventrodorsal view of the rabbit abdomen.

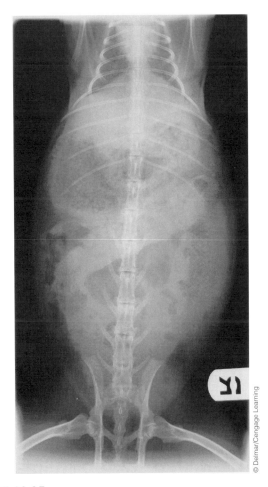

FIGURE 10-25

Ventrodorsal view of the rabbit abdomen.

Lateral View of the Rabbit Thorax

Positioning:

- Right lateral recumbency.
- Forelimbs are extended cranially and hindlimbs caudally; secure with tape.
- Sternum should be parallel to the cassette; this can be accomplished with the use of sponges.

Centering:

- Center on sternum.

Collimation:

- Slightly cranial of the thoracic inlet and slightly caudal of the last rib.

Labeling:

- R marker is placed just behind the forelegs.

Technique:

- Measure at the last rib.

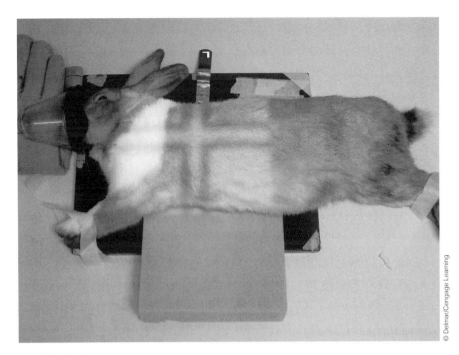

FIGURE 10-26

Proper positioning for lateral view of the rabbit thorax.

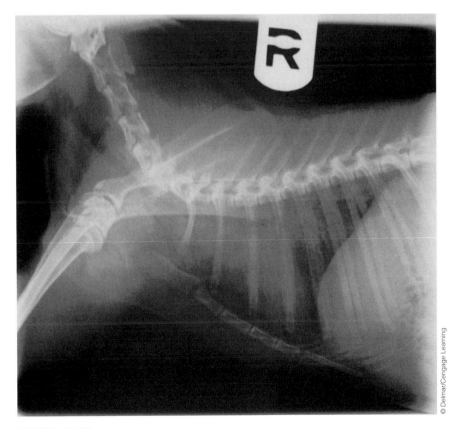

FIGURE 10-27

Lateral view of the rabbit thorax.

Ventrodorsal View of the Rabbit Thorax

Positioning:

- Dorsal recumbency.
- Forelimbs are extended cranially and hindlimbs caudally; secure with tape.
- Sternum should be placed directly over the spinal column.

Centering:

- Center on fourth intercostal space.

Collimation:

- Slightly cranial of the thoracic inlet and slightly caudal of the last rib.

Labeling:

- R marker is placed just behind the forelegs.

Technique:

- Measure at the last rib.

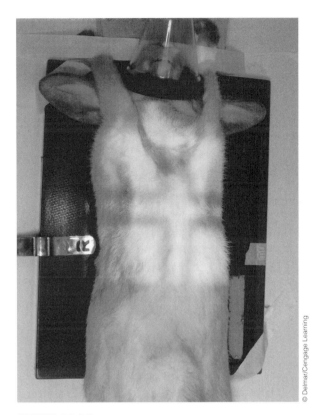

FIGURE 10-28

Proper positioning for ventrodorsal view of the rabbit thorax.

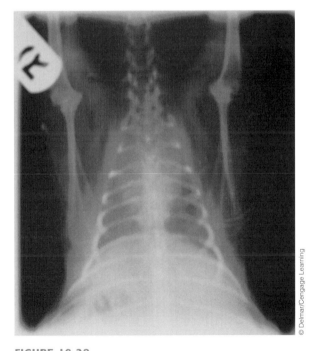

FIGURE 10-29

Ventrodorsal view of the rabbit thorax.

Lateral View of the Rabbit Skull

Positioning:

- Lateral recumbency.
- Use sponge wedges under the neck and nose to align to skull parallel to the cassette and prevent obliquity.
- Position the ears at the top of the skull, ensuring they are away from the area to be viewed.

Centering

- Midskull.

Collimation:

- Cranial of the nose and just caudal of the base of the skull.

Labeling:

- Just above the nose.

Technique:

- Thickest portion of the skull.

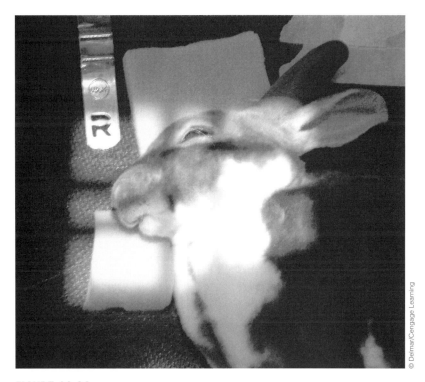

FIGURE 10-30

Proper positioning for lateral view of the rabbit skull.

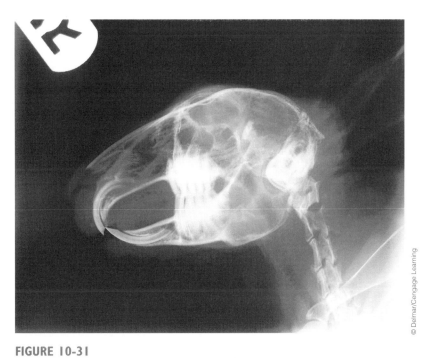

FIGURE 10-31

Lateral view of the rabbit skull.

Dorsoventral View of the Rabbit Skull

Positioning:

- Dorsoventral recumbency.
- Secure to head flat to the cassette with tape at the base of the skull.
- Position the ears to the sides, ensuring they are away from the viewing area.
- Ensure that the skull is parallel to the cassette.

Centering:

- Midskull.

Collimation:

- Cranial of the nose and just caudal of the base of the skull.

Labeling:

- Beside the nose.

Technique:

- Thickest portion of the skull.

FIGURE 10-32

Proper positioning for dorsoventral view of the rabbit skull.

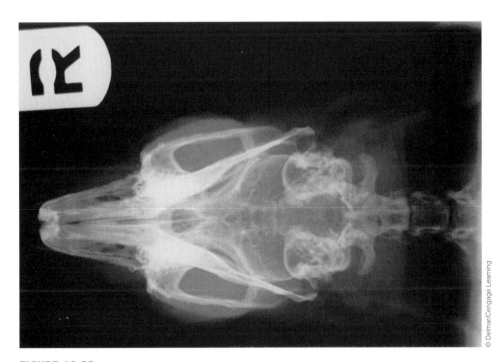

FIGURE 10-33

Dorsoventral view of the rabbit skull.

Lateral Oblique View of the Rabbit Skull

Positioning:

- Lateral recumbency (right or left).
- Sponge wedges are placed under the skull, tilting the head at a 45-degree angle to the cassette.
- Position the ears to the sides, ensuring they are away from the viewing area.

Centering:

- Midskull.

Collimation:

- Cranial of the nose and just caudal of the base of the skull.

Labeling:

- Beside the nose.

Technique:

- Thickest portion of the skull.

FIGURE 10-34

Proper positioning for lateral oblique view of the rabbit skull.

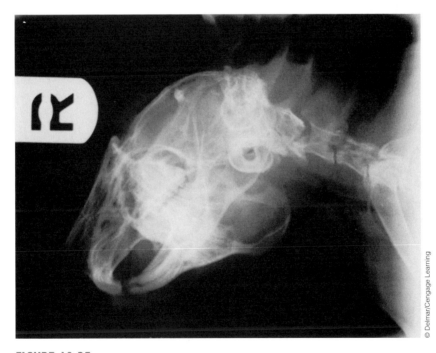

FIGURE 10-35

Lateral oblique view of the rabbit skull.

Lateral View of the Ferret Abdomen

Positioning:

- Right lateral recumbency.
- Forelimbs are extended cranially and hindlimbs caudally; secure with tape.
- Sternum should be parallel to the cassette; this can be accomplished with the use of sponges.

Centering:

- Center slightly caudal of the last rib.

Collimation:

- Slightly cranial of the xiphoid and slightly caudal of the pubis.

Labeling:

- R marker is placed in the inguinal area.

Technique:

- Measure at the last rib.

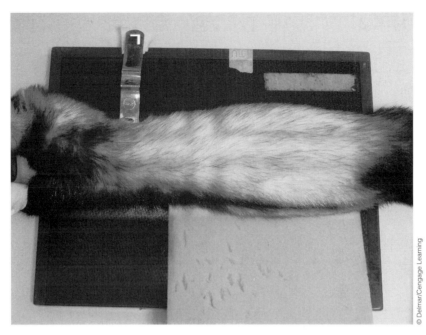

FIGURE 10-36

Proper positioning for lateral view of the ferret abdomen.

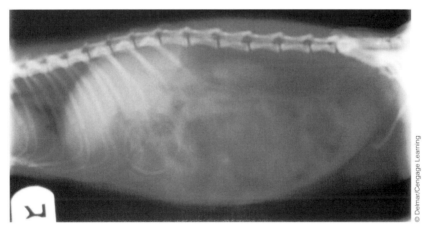

FIGURE 10-37

Lateral view of the ferret abdomen.

Ventrodorsal View of the Ferret Abdomen

Positioning:

- Dorsal recumbency.
- The front portion is secured with sandbags or tape to keep the trunk bilaterally symmetrical.
- The hind limbs are extended and secured with tape.

Centering:

- Center slightly caudal of the last rib.

Collimation:

- Slightly cranial of the xiphoid and slightly caudal of the pubis.

Labeling:

- R marker is placed in the inguinal area.

Technique:

- Measure at the last rib.

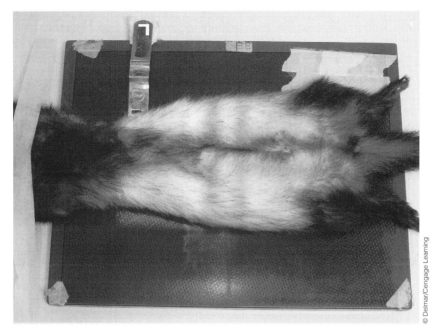

FIGURE 10-38

Proper positioning for ventrodorsal view of the ferret abdomen.

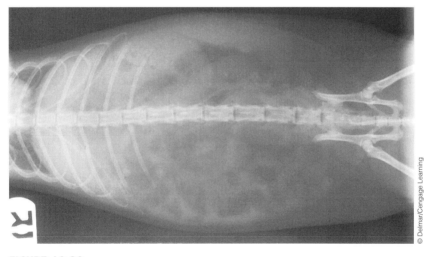

FIGURE 10-39

Ventrodorsal view of the ferret abdomen.

Lateral View of the Ferret Thorax

Positioning:

- Right lateral recumbency.
- Forelimbs are extended cranially and hindlimbs caudally; secure with tape.
- Sternum should be parallel to the cassette; this can be accomplished with the use of sponges.

Centering:

- Center on the xiphoid process.

Collimation:

- Slightly cranial of the thoracic inlet and slightly caudal of the dorsal aspect of the last rib.

Labeling:

- R marker is placed above the shoulders.

Technique:

- Measure at the last rib.

© Delmar/Cengage Learning

FIGURE 10-40

Proper positioning for lateral view of the ferret thorax.

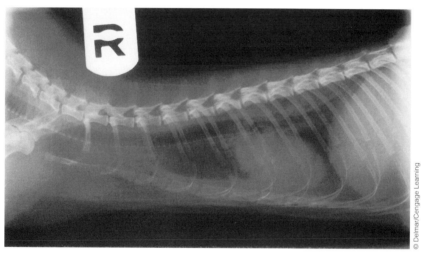

© Delmar/Cengage Learning

FIGURE 10-41

Lateral view of the ferret thorax.

Ventrodorsal View of the Ferret Thorax

Positioning:

- Dorsal recumbency.
- The front portion is secured with sandbags or tape to keep the trunk bilaterally symmetrical.
- The hind limbs are extended and secured with tape.

Centering:

- Center on the xiphoid process.

Collimation:

- Slightly cranial of the thoracic inlet and slightly caudal of the dorsal aspect of the last rib.

Labeling:

- R marker is placed above the shoulders.

Technique:

- Measure at the last rib.

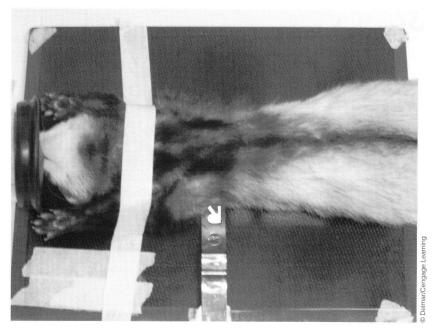

FIGURE 10-42

Proper positioning for ventrodorsal view of the ferret thorax.

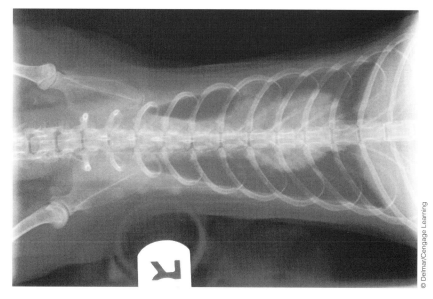

FIGURE 10-43

Ventrodorsal view of the ferret thorax.

Lateral Whole Body View of the Ferret

Positioning:

- Right lateral recumbency.
- Forelimbs are extended cranially and hindlimbs caudally; secure with tape.
- Sternum should be parallel to the cassette; this can be accomplished with the use of sponges.

Centering:

- Halfway between the dorsal aspect of the last rib and the xiphoid process.

Collimation:

- Slightly cranial of the thoracic inlet and slightly caudal of the pubis.

Labeling:

- R marker is placed in the inguinal region.

Technique:

- Measure at the last rib.

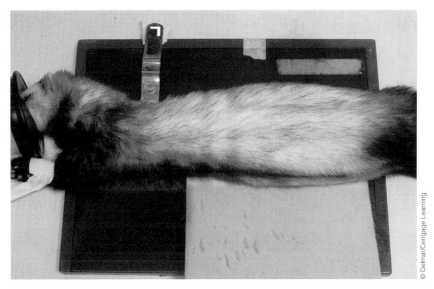

FIGURE 10-44

Proper positioning for lateral whole body view of the ferret.

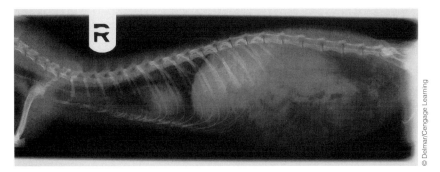

FIGURE 10-45

Lateral whole body view of the ferret.

Ventrodorsal Whole Body View of the Ferret

Positioning:

- Dorsal recumbency.
- The front portion is secured with sandbags or tape to keep the trunk bilaterally symmetrical.
- The hind limbs are extended and secured with tape.

Centering:

- Halfway between the dorsal aspect of the last rib and the xiphoid process.

Collimation:

- Slightly cranial of the thoracic inlet and slightly caudal of the pubis.

Labeling:

- R marker is placed in the inguinal region.

Technique:

- Measure at the last rib.

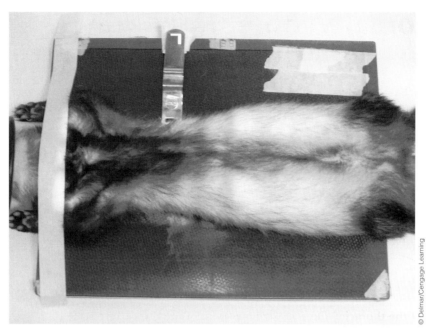

FIGURE 10-46

Proper positioning for ventrodorsal whole body view of the ferret.

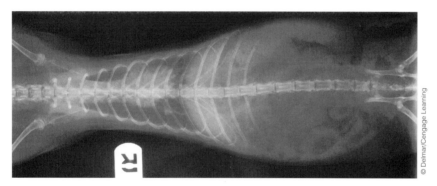

FIGURE 10-47

Ventrodorsal whole body view of the ferret.

Index

6.0 FEDERAL GOVERNMENT CLIENTS

6.1 Except as expressly authorized by Cengage Learning, Federal Government clients obtain only the rights specified in this Agreement and no other rights. The Government acknowledges that (i) all software and related documentation incorporated in the Licensed Content is existing commercial computer software within the meaning of FAR 27.405(b)(2); and (2) all other data delivered in whatever form, is limited rights data within the meaning of FAR 27.401. The restrictions in this section are acceptable as consistent with the Government's need for software and other data under this Agreement.

7.0 DISCLAIMER OF WARRANTIES AND LIABILITIES

7.1 Although Cengage Learning believes the Licensed Content to be reliable, Cengage Learning does not guarantee or warrant (i) any information or materials contained in or produced by the Licensed Content, (ii) the accuracy, completeness or reliability of the Licensed Content, or (iii) that the Licensed Content is free from errors or other material defects. THE LICENSED PRODUCT IS PROVIDED "AS IS," WITHOUT ANY WARRANTY OF ANY KIND AND CENGAGE LEARNING DISCLAIMS ANY AND ALL WARRANTIES, EXPRESSED OR IMPLIED, INCLUDING, WITHOUT LIMITATION, WARRANTIES OF MERCHANTABILITY OR FITNESS FOR A PARTICULAR PURPOSE. IN NO EVENT SHALL CENGAGE LEARNING BE LIABLE FOR: INDIRECT, SPECIAL, PUNITIVE OR CONSEQUENTIAL DAMAGES INCLUDING FOR LOST PROFITS, LOST DATA, OR OTHERWISE. IN NO EVENT SHALL CENGAGE LEARNING'S AGGREGATE LIABILITY HEREUNDER, WHETHER ARISING IN CONTRACT, TORT, STRICT LIABILITY OR OTHERWISE, EXCEED THE AMOUNT OF FEES PAID BY THE END USER HEREUNDER FOR THE LICENSE OF THE LICENSED CONTENT.

8.0 GENERAL

8.1 *Entire Agreement.* This Agreement shall constitute the entire Agreement between the Parties and supercedes all prior Agreements and understandings oral or written relating to the subject matter hereof.

8.2 *Enhancements/Modifications of Licensed Content.* From time to time, and in Cengage Learning's sole discretion, Cengage Learning may advise the End User of updates, upgrades, enhancements and/or improvements to the Licensed Content, and may permit the End User to access and use, subject to the terms and conditions of this Agreement, such modifications, upon payment of prices as may be established by Cengage Learning.

8.3 *No Export.* The End User shall use the Licensed Content solely in the United States and shall not transfer or export, directly or indirectly, the Licensed Content outside the United States.

8.4 *Severability.* If any provision of this Agreement is invalid, illegal, or unenforceable under any applicable statute or rule of law, the provision shall be deemed omitted to the extent that it is invalid, illegal, or unenforceable. In such a case, the remainder of the Agreement shall be construed in a manner as to give greatest effect to the original intention of the parties hereto.

8.5 *Waiver.* The waiver of any right or failure of either party to exercise in any respect any right provided in this Agreement in any instance shall not be deemed to be a waiver of such right in the future or a waiver of any other right under this Agreement.

8.6 *Choice of Law/Venue.* This Agreement shall be interpreted, construed, and governed by and in accordance with the laws of the State of New York, applicable to contracts executed and to be wholly preformed therein, without regard to its principles governing conflicts of law. Each party agrees that any proceeding arising out of or relating to this Agreement or the breach or threatened breach of this Agreement may be commenced and prosecuted in a court in the State and County of New York. Each party consents and submits to the nonexclusive personal jurisdiction of any court in the State and County of New York in respect of any such proceeding.

8.7 *Acknowledgment.* By opening this package and/or by accessing the Licensed Content on this Web site, THE END USER ACKNOWLEDGES THAT IT HAS READ THIS AGREEMENT, UNDERSTANDS IT, AND AGREES TO BE BOUND BY ITS TERMS AND CONDITIONS. IF YOU DO NOT ACCEPT THESE TERMS AND CONDITIONS, YOU MUST NOT ACCESS THE LICENSED CONTENT AND RETURN THE LICENSED PRODUCT TO CENGAGE LEARNING (WITHIN 30 CALENDAR DAYS OF THE END USER'S PURCHASE) WITH PROOF OF PAYMENT ACCEPTABLE TO CENGAGE LEARNING, FOR A CREDIT OR A REFUND. Should the End User have any questions/comments regarding this Agreement, please contact Cengage Learning at Delmar.help@cengage.com.

System Requirements for: Back of Book (BOB) Flash Shell CD/DVD-ROMs

Minimum System Requirements:

PC:

- Operating System: Windows 2000 w/ SP4, XP w/ SP2, Vista
- Hard Drive space: 200MB
- Screen resolution: 1024 x 768 pixels
- 8x CD-ROM or DVD-ROM drive
- Sound card and listening device required for audio features
- An Internet connection, Firefox 2 or Internet Explorer 6 & 7 for Internet based content
- Microsoft® Word is required to edit the Instructor's Manual and Microsoft PowerPoint® is required to edit the presentations

Mac:

- Operating System: Mac OS X 10.4 and 10.5
- Microsoft Office® 2004 or greater (for viewing files)
- Hard Drive space: 200 MB
- Minimum Screen resolution: 1024 x 768 pixels
- 8x CD-ROM or DVD-ROM drive
- Sound card and listening device required for audio features

- An Internet connection, Firefox 2 or Safari 3 for Internet based content

PC Setup Instructions:

1. Insert disc into CD-ROM drive. The program should start automatically. If it does not, go to step 2.
2. From My Computer, double-click the icon for the CD drive.
3. Double-click the *start.exe* file to start the program.

Mac Setup Instructions

1. Insert disc into CD-ROM drive. The Cengage icon will pop up on your desktop and a window should pop up with a file called *start*. If it does not, go to step 3.
2. Double-click the *start* file, to start the program.
3. Click on the Cengage icon on your desktop. Go to Step 2.

Technical Support:
Telephone: 1-800-648-7450
8:30 AM - 6:30 PM Eastern Time
E-mail: delmar.help@cengage.com

Microsoft®, Microsoft Office®, Microsoft PowerPoint®, Microsoft Word®, Windows®, Windows XP® and Windows Vista® are trademarks of the Microsoft Corporation.
Mac® and Mac OS X® are trademarks of Apple Inc., registered in the U.S. and other countries.